LOST ANIMALS

LOST ANIMALS

EXTINCTION AND THE PHOTOGRAPHIC RECORD

Errol Fuller

PRINCETON UNIVERSITY PRESS
PRINCETON AND OXFORD

Published in the United States, Canada, and the Philippine Islands in 2014 by
Princeton University Press, 41 William Street, Princeton, New Jersey 08540
nathist.press.princeton.edu

Published in 2013 by Bloomsbury Publishing Plc, 50 Bedford Square, London WC1B 3DP

Library of Congress Control Number 2013946564
ISBN 978-0-691-16137-2

This book is produced using paper that is made from wood grown in managed sustainable forests.
It is natural, renewable and recyclable. The logging and manufacturing processes conform to the
environmental regulation of the country of origin.

Commissioning Editor: Jim Martin
Designed by Errol Fuller

Printed in China
10 9 8 7 6 5 4 3 2 1

Endpapers: Female Quagga, photographed by Frederick York during the summer of 1870
at the London Zoo.
Page 2: Ivory-billed Woodpecker on the head of J. J. Kuhn, photographed by James
Tanner on 6th March 1938. Courtesy of Nancy Tanner.
Facing page: New Zealand Bush Wren, photographed by Herbert Guthrie-Smith in 1913.

ITEM CHARGED

Patron: HANNAH GABRIELLE SHELAN

Patron Barc

Patron Gro UNDERGRAD

Due Date: 2/26/2019

Title: Lost animals : extinction and the photographic record / Errol Fuller.

Author: Fuller, Errol,

Call Numbe 591.68 F9585lo

Enumeratio

Chronology:

Copy: 1

Item Barcod

For Tessie

A flock of ten Pink-headed Ducks at Foxwarren Park, Surrey, England in 1929 (also reproduced on page 37). The name of the photographer is uncertain although it may have been a well-known specialist in Indian birds named Salim Ali. Courtesy of Frank S. Todd.

Contents

Introduction

Some years ago I wrote a book called *Extinct Birds.* It was packed with illustrations, mostly reproductions of paintings that showed the featured birds, some of them by truly great artists. Then there were a few grainy photographs, taken when some species were still extant.

When friends or acquaintances thumbed through the book a peculiar and unexpected thing became very noticeable. They were attracted by the high quality of the paintings, of course, but they were truly riveted by the photos. They would pause over them and just gaze, sometimes even raising the book towards their eyes in the vain hope that this action would allow them to see more – more than there really was to see! Almost always the same question cropped up. 'Is this real or have you just faked it?'

All this was despite the fact that most of these images were, inevitably, poor in quality (for many were taken in difficult circumstances in the early days of photography), and showed little detail.

It seems that a photograph of something lost or gone has a power all of its own, even though it may be tantalisingly inadequate.

And this is one of the reasons for the present book. Here the reader will find images that are far less than satisfactory in quality. Many are from days when only black-and-white photography was possible, and in an age when we have become so accustomed to colour there are those who will find this disappointing. But despite all of the handicaps, these photos are evocative and moving records of creatures that are gone. They are close enough to touch – almost, but not quite!

To compensate in some small way for any frustration the interested reader may feel, and to provide a semblance of completeness in his or her mind, a gallery of paintings of most of the featured animals is included as an appendix at the back of the book. Some, the Caribbean Monk Seal for instance, are omitted from the appendix as it is felt that such an illustration would serve little purpose. Others, like the Guam Flycatcher, are also left out as they are clearly depicted in the existing photos. Those paintings that are included may help anyone struggling with just a black-and-white or blurred image to visualise more exactly what the mammal or bird actually looked like.

Certain factors should be borne in mind when viewing the photographs. They may have been taken in exceptionally difficult circumstances, when only a fleeting view of the subject was obtained. And, of course, in these days of digital technology, it is easy to forget what a complex and expensive process photography once was. Cumbersome, heavy equipment was needed, and this often had to be hauled for miles over difficult terrain. Lighting was critical; so too was complete stillness of the subject. Wet plates that dried out before the subject was properly in position and the need to be relatively close to it were problems that had to be overcome. Also, there was no way of knowing just what the camera had caught. Immediate inspection (something we take for granted today) was out of the question. Film had to be 'developed' in a dark room, and this facility was often miles (or perhaps many days travel) from the place where the photos were taken. Also significant is the fact that photographers often had no idea how important their photos would become. They didn't necessarily have any insight that their subject would soon become extinct.

Even though some of the pictures reproduced in this book are particularly poor in quality, no attempt has been made to tamper or enhance them with the various modern techniques now available. They are shown here as they exist and must speak for themselves. The photograph of the Mamo on page 155, for instance, is hardly a masterpiece of the photographer's art, but in its own way it is full of atmosphere, even poignancy.

Anyone who is inclined to disappointment must bear in mind all of these factors when looking at the images. Hopefully the viewer will accept them for what they are, and for what they represent.

The circumstances under which photos were taken are often interesting or curious, although sometimes, of course, little or nothing is known. As might be expected, several were taken in zoos, but the majority were shot in other situations.

Sometimes it is surprisingly difficult to discover who actually took a surviving photograph, or where and when it was shot. Reference books or websites on the internet may be valuable tools, but they sometimes promote conflicting information. The internet is particularly open to misleading claims. People or organisations often suggest a photograph is theirs when in fact they have no genuine claim to it. The picture of Bachman's Warbler featured on page 137, for instance, occurs on the internet under a variety of names, all of whom appear willing to take some kind of credit for it. The truth seems to be that it was taken by a gentleman named J. H. Dick in 1958, but this is by no means certain.

In any work on extinct creatures, the problem of what is extinct and what is not always rears its head. For a number of species, hopes are regularly expressed that individuals may still survive in some out-of-the-way place. The Thylacine, the Pink-headed Duck, and the Paradise Parrot are all examples of this kind of wishful thinking. Perhaps they do still survive, but it is far, far more likely that they do not. Much attention was recently focused on the supposed survival of the Ivory-billed Woodpecker, but the claims came (as many people suspected they would) to nothing.

The question of the borderline between species and subspecies is another that causes controversy. In this book creatures that are generally regarded as extinct races of still-extant species have not been included.

(*Above*). Extinct or not? The Thylacine is probably extinct, but rumours persist that individuals are still being seen, even though the species has been classified as extinct since 1936. These two were photographed at a zoo in Washington during the first decade of the twentieth century. The photographer is unknown.

(*Above*). Another species for which hopes of survival are still frequently expressed is the Pink-headed Duck. These hopes are probably forlorn, but expeditions are often launched with the intention of finding an existing colony in some remote spot. This extraordinary photo of ten individuals was taken in 1929 at Foxwarren Park, Surrey, England, where they were being kept among a collection of ornamental waterfowl. At this time the species may well have been extinct in its homeland in India and other parts of Asia. The photographer is unknown, but it was possibly the well-known Indian bird specialist Salim Ali. Photograph courtesy of Frank S. Todd.

There are two notable exceptions to this principle. These are the Quagga and the Heath Hen, both of which are generally regarded as races of species that otherwise still exist. They are included because they have gained such a clear identity in the minds of those who have interested themselves in the subject of extinct animals. And since reasonably good photographs exist it would be something of a shame to leave them out. If this lack of consistency offends, then so be it …

Featured animals are restricted to birds and mammals. To include reptiles, amphibians, fish, invertebrates or plants would be somewhat problematic, and in any case would be a large enough subject for another book.

There are glaring gaps in the photographic record, and many birds and mammals one might hope to see remain unrepresented. Photographs of extinct mammals are in particularly short supply. Birds seem to have attracted photographers to a much greater extent, although the most famous of all extinct birds – the Dodo (*Raphus cucullatus*) – was never caught on camera; it vanished at far too early a date. Perhaps the next most famous is the Great Auk (*Alca impennis*), and this species almost survived into what might be termed 'photographic times' – but not quite. The last known pair were killed on an Icelandic island (Eldey) during 1844, the same year in which William Fox Talbot (1800–1877) began to publish the first photographic book, *The Pencil of Nature.*

It is surprising to find that other more recently extinct creatures were not caught on camera. New Zealand's Huia (*Heteralocha acutirostris*), for instance, didn't become extinct until the early years of the twentieth century, and several pairs were held in captivity. There are even drawings of one at London Zoo; but there are no photos – at least there are none that are known.

The same could be said of many other extinct creatures. This book contains what is available.

Here it might be useful to issue a small warning. There are many photographic images posted on the internet that purport to show extinct species in life. In fact these often show something quite different. Usually errors of this kind are based on misunderstanding or a lack of experience in the subject, rather than any real intent to deceive. Sometimes, of course, the deception is deliberate.

(*Above*). One of a series of photos taken by James Tanner on March 6th, 1938 showing a young Ivory-billed Woodpecker on the sleeve of his colleague J. J. Kuhn. This particular photograph and several others from the series were discovered in 2009 by Tanner's widow Nancy, and Stephen Lyn Bales who wrote a book, *Ghost Birds* (2010), about the species and Tanner's attempt to save it. Courtesy of Nancy Tanner.

Although for some species (the Thylacine, for instance) many photos exist, in most cases the actual photographs shown here represent all that we have. There may be others, but any such have proved impossible to trace. As the case of the film of the Imperial Woodpecker and several recently discovered archival photos of the Ivory-billed proves, more of the truth may still be out there!

(*Above*). A Laysan Rail on its nest. Taken by Walter K. Fisher in May 1902, this photograph is used by permission of Denver Museum of Nature and Science. The species became extinct as World War II drew to a close.

The crux of the matter ... is not who or what kills the last individual. That final death reflects only a proximate cause. The ultimate cause, or causes, may be quite different. By the time the death of the last individual becomes imminent, a species has already lost too many battles in the war of survival ... Its evolutionary adaptability is largely gone. Ecologically it has become moribund. Sheer chance, among other factors, is working against it. The toilet of destiny has been flushed.

David Quammen

Atitlán Giant Grebe
Podilymbus gigas

Sometimes the extinction of a species can be traced back to a single cause. More often there are a number of contributory factors. But the case of the Giant Grebe of Lake Atitlán has everything: murder, habitat destruction, political interference, the introduction of an alien species, dilution of the bloodstock by hybridisation, the effects of tourism, pollution, civil war, and an earthquake.

Perhaps just as extraordinary as any of these things is the fact that the final years of its destruction were chronicled in the most intense detail by a fanatically dedicated woman named Anne LaBastille (1935–2011). In numerous magazine articles and papers in learned journals she recounted the story of her attempts to save the birds from disaster, and pleaded for help with her efforts. Finally, when all was lost, she wrote a highly personal book. Called *Mama Poc* (*poc* being the sound the birds made) and published in 1990, it is a thorough record of her forlorn struggle to save the species.

Although often referred to as a 'giant', by the standards of many birds it could hardly be called that. At around 50 centimetres long (20 inches) it was only a giant in comparison to most other kinds of grebe. In fact this was a large, flightless form that had evolved from the much more

(*Facing page*). One of several photos of the Atitlán Giant Grebe taken by photographer David G. Allen.

widespread and smaller Pied-billed Grebe (*Podilymbus podiceps*). Long ago a group of Pied-billed Grebes chanced on Lake Atitlán, stayed, and over millennia lost the power of flight and grew much larger. Perhaps their greater size allowed them to stay underwater for longer and take better advantage of whatever food resources this particular lake offered. Why this happened is unknown, but either way, the 'giant' grebe evolved here and, seemingly, nowhere else.

(*Above*). Anne LaBastille with a an emaciated juvenile that she tried to nurse back to health.

The species was first described in 1929, but almost nothing was known of it until Ms LaBastille visited Atitlán during 1965. Five years previously, a survey had concluded that the population consisted of 200 to 300 birds but, though small in number, this population was considered stable. But in the five years between the time of the survey and Anne LaBastille's arrival this 'stable' population plummeted. By 1965 there were only around 80 birds left. One immediate reason was easy enough to spot: the local human population was cutting down the reed beds (in which the species nested) at a furious rate. This destruction was driven by the needs of a fast growing mat-making industry. But there were other problems. Pan Am, the now-defunct American airline, was intent on developing the lake as a tourist destination for fishermen. However, there was a major problem with this idea; the lake lacked any suitable sporting fish! To compensate for this rather glaring defect, a specially selected species of fish was introduced. Even the name of this fish gives enough clues to suggest a nasty outcome; it is called the Large-mouthed Bass (*Micropterus salmoides*), and the introduced individuals immediately turned their attentions to the crabs and small fish that lived in the lake, thus competing with the few remaining grebes for food. There is also little doubt that they sometimes gobbled up the zebra-striped grebe chicks.

Anne LaBastille somehow managed to overcome these enormous difficulties. Instead of returning to her home in the United States she stayed, raising the profile of the species and also raising funds for her work. While doing this she established a refuge on the lake's banks. Around this time she was able to interest the *National Geographic* in her project, and they sent an independent photographer, David G. Allen, to take pictures of the birds.

To give him a proper chance of getting good shots it was arranged that Mr Allen would stay for a month. He required a high tower with a platform and blind in order to do his work and this was constructed from spliced wood with bamboo supports. It was positioned close to a spot in which a pair of grebes had nested and just laid eggs. The photographer's dedication to his task impressed Ms. LaBastille enormously. She later wrote:

> *Once the tower, platform and blind were installed we saw very little of David. Armando* [Anne LaBastille's helper] *took him across the lake by boat each dawn, escorted him right into the blind, and left. That way, if the pocs were watching, they thought both intruders had departed. Birds cannot count. The only way we knew whether David was alive or dead was the thin trickle of cigarette smoke issuing from the top of the blind. He was seldom without a cigarette in his mouth. In the late afternoon Armando picked him up again … I had never seen a photographer work as hard as David Allen, and he gained my total respect.*

Curiously, the photos he obtained were never used in *National Geographic* itself, despite their quality and the intrinsic interest of the subject matter.

Within eight years of Anne LaBastille's arrival, the population had climbed to 200, but new threats were on the

(*Facing page*). One of David G. Allen's photographs of the nest that his hide and tower overlooked, showing two eggs and a newly hatched chick.

(*Facing page*). A female Atitlán Grebe surfacing from a dive with food, photographed by David G. Allen.

horizon. Whereas the indigenous human population had compromised over destruction of the reed beds, newcomers from Guatemala City were less helpful. They were building homes at the lakeside and ripping out any vegetation that stood between their houses and the shoreline. Another new arrival was an avian one. The new conditions and the fragile situation of their giant relatives provided an opportunity for any Pied-billed Grebes that happened to stray into the area, and this species – with its ability to fly – was much more robust and vigorous than the Atitlán Grebes. Not only did it compete for food, but individuals began to hybridise with the resident population, and started a process of breeding them out of existence.

During the mid-1970s a natural disaster that led to the deaths of some 22,000 humans had another, lesser known, consequence. This disaster was an earthquake and it caused the lowering of the lake's surface by 6 metres (20 feet). The grebe refuge was left high and dry and it became necessary to transplant thousands of reed clumps. A man named Edgar Bauer assisted in this operation and became an enormous help with the project to save the grebes, but he was gunned down by unknown assassins. Ms LaBastille lost her greatest supporter just when she needed him most. The political situation in Guatemala was so tense at that time that no-one dared step forward to replace him.

By 1980 just 50 Atitlán Grebes were left. Three years later the number was down to 32. A last-ditch attempt to save them became necessary, and the remaining birds were rounded up. But as this was being done a dozen of them simply rose into the air and flew off. They were no longer pure Atitlán Grebes, but hybrids that were able to fly.

In 1978 it was announced that although a few pure individuals might still survive the species was biologically – if not actually – extinct.

Ms. LaBastille wrote:

There will be no need ever to run a census again on Lake Atitlán.

(*Above*). A male Atitlán Grebe, photographed by David G. Allen.

Alaotra Grebe
Tachybaptus rufolavatus

Although the Alaotra Grebe became extinct quite recently only one photograph seems to exist – and even this one is not very clear. However, in many respects it is remarkable that it exists at all, and it is only through the efforts of a man called Paul Thompson that it does.

As far as is known this smallish grebe (around 25 centimetres or 10 inches in length) was restricted to Lake Alaotra and a few nearby lakes in north-east Madagascar. Few naturalists ever saw it, and information concerning the species is very sparse. For decades the great island of Madagascar has been becoming something of an ecological disaster and this grebe is only one of many victims.

The photo was taken in the September of 1985. Paul Thompson was taking part in an expedition sponsored by a wildlife protection agency, and during the course of his journey found himself at Lake Alaotra. These are the circumstances, in his own words, of how he came to take the picture:

> *On our first day there we took a small pirogue* [a small flat-bottomed boat] *out with a couple of lads, but it was very leaky and we were lucky not to sink. So next day I didn't take the camera … but this time we were in a vastly superior pirogue … and I missed plenty of good photo opportunities … On our last morning there we took out a reasonable pirogue but Dave* [Paul's companion on the trip] *got sick and I took photos in the limited*

(*Above*). Paul Thompson's photograph of the Alaotra Grebe. Taken during 1985 in rather difficult circumstances, it seems to be the only one in existence. After this time only hybrid birds could be found, but even these are now gone. Courtesy of Paul Thompson and BirdLife International.

*time we had. We noted all three species of grebe
[that occurred there] and some we thought were
hybrids of part Alaotra parentage. I gave the
rights for use [of photos] to BirdLife.*

Paul thinks he may have taken more than one picture of
the grebe, but many years have passed and his memories of the
matter have faded; there is no trace of another in the
BirdLife archives. He returned to the lake as 1987 was passing
into 1988 but failed to find any more birds. Not only is he the
only person to have photographed the Alaotra Grebe, he may
also be the last ornithologist to have seen it; a few later
sightings are thought to have been observations of hybrid
birds. Like the Giant Grebe of Lake Atitlán, this grebe also
hybridised with related species (in this case the Little Grebe
Tachybaptus ruficollis) until its blood-line was almost
completely overwhelmed.

Hybridisation is clearly something that occurs quite
often among birds in the grebe family. There are more than
20 different species and some are very closely related.
Occasionally, a species of grebe will colonise an isolated lake
and in this secluded place will, over time, develop features
that separate it from its ancestral stock. Then, if a new wave
of the original stock invades the area the two forms will
hybridise. The genetic pool of the new wave is usually
more vibrant and eventually the first colonisers are swamped
by the newcomers. This seems to be one of the causes of the
extinction of the Alaotra Grebe, although it was certainly not
the only one. As mentioned in his account, Paul Thompson
himself noticed hybrid grebes on the lake, but he was certain
that he also saw pure bred birds.

Overfishing – which limited food supplies – and the equipment fishermen used – in which the birds were liable to get tangled – were other causes, as well as environmental deterioration and an introduced carnivorous fish. In these ways the extinction of this species parallels that of the Atitlán Giant Grebe. Another contribution to the species' plight was the fact that evolution had resulted in rather small wings, and flight was therefore poor. This meant that individuals were unable to disperse to more favourable areas when Lake Alaotra became so unsuitable.

Although it is usual not to declare animals extinct until several decades have passed since the last confirmed sighting, an exception has been made in this case. After a number of recent expeditions to the area and evaluation of the evidence, Leon Bennun of BirdLife International, an organisation devoted to the plight of endangered birds, declared:

No hope now remains for this species.

Pink-headed Duck
Rhodonessa caryophyllacea

Pink is a most unusual colour in birds, and perhaps this is one of the main reasons why a duck with a strikingly pink head and neck is still avidly sought after by those who hope to re-discover extinct species. More than three quarters of a century have passed since Pink-headed Ducks were reliably observed in the wild, yet expeditions to find them again are still launched with some regularity. None have yet proved successful, but perhaps the search is not entirely forlorn.

During the 19th century, the Pink-headed Duck was an inhabitant of the swamps, rivers and reed beds that studded the plains surrounding the lower reaches of the Ganges and the Brahmaputra Rivers. It was not common, but neither was it particularly rare. The lowlands stretched far and wide on either side of these mighty rivers and due to the difficulty of the terrain – with huge areas of marsh dissected by streams and water-channels – large tracts of the area remained impenetrable and unexplored. Much has changed now. The Tiger-infested jungles and swamps of a land once known as Bengal are now part of the country more regularly known as Bangladesh. Wetlands have been drained and brought under some degree of control (although disastrous floods do still occur), and settlement and cultivation plus an ever-increasing population have pushed back the wilderness. Parts of the area are now among the most densely populated places on earth. Somewhere in the middle of all this development and change, the Pink-headed Duck quietly disappeared.

(*Above*). During the 1920s and 1930s a number of Pink-headed Ducks were kept in captivity with other waterfowl at Foxwarren Park, Surrey, England, far from their natural home in India and Burma. Here at Foxwarren, in 1926, they were photographed by David Seth-Smith, a man who took many pictures of zoo animals as well as producing paintings of wildlife and writing books and articles for journals on natural history and related subjects. This picture shows a male (left) and a female. The characteristic straight-necked posture of the species and the rather peculiar shape of its head is clearly revealed. It is quite possible that some of these captive birds outlived all of their wild counterparts.

(*Above and facing page*). Two photos of Pink-headed Ducks taken at Foxwarren Park, Surrey in 1926 by David Seth-Smith.

Always shy and wary, the species was difficult to flush from cover, so, unlike many duck species, these were not classic sporting birds. They weren't even particularly sought after for the table, at least as far as British colonists were concerned, although any such delicacy is unlikely to have deterred the hungry local population. What did attract British residents seems to have been the peculiarity of the head and neck colouring, and the curiosity it aroused.

Although still frequently encountered, a decline seems to have set in during the last quarter of the 19th century, and once it began it progressed at a rapid pace. In the first decade of the 20th century the duck's increasing scarcity was being noticed, and by the 1920s the species was virtually extinct. The last definite sighting in the wild appears to have been of a bird seen in the mid 1930s.

Reasons for the duck's extinction are unknown. Over-hunting does not seem to have been severe enough to have been a vital factor, and at the time of the species' decline there were still vast tracts of unspoiled country – in fact there still are. In addition to Bengal, the species also inhabited Assam in north-east India and parts of Burma. There must have existed some entirely unrecognised factor that affected the duck in such a drastic way.

Considering that the birds were virtually gone by the 1920s, it is surprising to find that a consignment of three living pairs was received by Alfred Ezra (1872–1955) in southern England during 1926. Ezra, who got them from his brother in Calcutta, had a particular interest in Indian birds and kept a magnificent collection of waterfowl at his home, Foxwarren Park in Surrey. Then, even more amazingly, he received another ten birds three years later, four of which he sent to his friend Jean Delacour (1890–1985), a French waterfowl enthusiast. Where did they come from? It seems not to have been noted down at the time, and nobody is now alive who might remember.

Although thriving as individuals, the ducks at Foxwarren Park failed to breed and, apparently, made no attempt to do so. Neither did the ones kept by Delacour at his home in Clères, France. This may have been because they were kept with other waterbirds, a factor that perhaps discouraged breeding.

During their residence at Foxwarren, David Seth-Smith (1875–1963), a man greatly interested in zoos and their collections, saw the ducks and in 1926 he took several photographs. Although only in black-and-white, and therefore failing to reveal the striking pink of the head, the best of these shows the ducks with some clarity.

Gradually, all of Ezra's captive ducks died off and so too did their counterparts in France, but the date of the death of the last one remains something of a mystery. The year 1936 is probably correct. However, some authorities give a date of 1939 and others suggest 1945. Who knows which date is correct? But, of course, it no longer really matters.

(*Above*). The flock of ten Pink-headed Ducks at Foxwarren Park in 1929. The photographer is uncertain although it may have been the well-known Indian bird specialist Salim Ali. This photo is also featured on pages 6 and 7. In recent years it (and another similar picture) has been subjected to various modern adjustments and colouring processes in an attempt to emphasise the pink of the head. Courtesy of Frank S. Todd.

Heath Hen

Tympanuchus cupido cupido

(*Above*). The last Heath Hen, named 'Booming Ben' by some of his admirers, photographed on Martha's Vineyard around 1930 by Alfred O. Gross.

There is no doubt that the Heath Hen is a race of an otherwise still extant species, the Prairie Chicken. Nevertheless, like the Quagga (see pages 194–205), it has acquired such a clear identity in zoological literature and

history that leaving it out would seem something of an omission. Indeed, it is not unrealistic to say that it is one of the best known of all extinct birds. This is due, at least in part, to the fact that it was an inhabitant of the eastern seaboard of the United States, but it is also because its story is a dramatic one, and photographs exist of an individual that was known to be

(*Above*). This photograph of another individual was taken by Gross in the spring of 1924, and shows what he described as a 'characteristic attitude of repose.'

the last of its kind. These were taken by Alfred O. Gross, an enthusiast who fought long and hard to save the last few birds.

The species, *Tympanuchus cupido*, was once widely distributed across the plains of North America, and ornithologists divided it into several races. Three of these lived in the west, but the fourth, *cupido*, inhabited areas near to the Atlantic seaboard. The western birds became known as

Prairie Chickens, and those in the east were called Heath Hens. Presumably because of the growing human population in the

(*Above*). A second photograph taken by Gross during the spring of 1924.

(*Facing page*). Six more photos by Gross. These were taken on March 28th and 29th 1927. They were published together in his monograph *The Heath Hen*, produced for The Boston Society of Natural History (1928).

41

east, Heath Hens soon lost much of the habitat they required and, of course, being plump, juicy and available, they were hunted. A habit of congregating in open country made them particularly vulnerable to the hunter's gun and in addition to this danger they proved susceptible to diseases transmitted by domestic chickens.

(*Above*). A male Heath Hen displaying, photographed at an unknown date by George W. Field.

At some time during the 19th century Heath Hens disappeared entirely from mainland North America. But they weren't quite extinct.

A small population still lived on an island off the Massachusetts coast just south of Cape Cod. Known as Martha's Vineyard, the name itself is something of a mystery; nobody really knows who Martha was. It is one of the oldest English place-names in the Americas, having been christened early in the 17th century by the explorer Bartholomew Gosnold (1572–1607). It is thought that Martha may have been his daughter, who had died at an early age.

(*Above*). Another of Gross's pictures of the last Heath Hen as it wandered near to his hide in 1929.

On Martha's Vineyard around 100 birds survived, and this tiny relict population aroused the interest and support of various conservation-minded bodies. Due to their efforts the population rose and rose, until it seemed quite secure. By early 1916 there may have been as many as 2,000 birds.

But then disaster struck. On May 12th 1916, fire swept through the main breeding area. Many birds were killed and the conservationists' work was essentially ruined. The majority of the survivors were males, but even so the birds seemed to rally and numbers increased a little. Sadly, the improvement proved temporary. With few females the breeding pool was limited. Many birds succumbed to a mysterious disease, perhaps exacerbated by the extensive in-breeding. Numbers dropped rapidly and by the start of

1929 only a single individual – a male, christened
'Booming Ben' by his admirers – remained. Alfred Gross
took his portrait several times, but he was last seen on March
11th 1932.

(*Facing page*). A third photograph by Alfred O. Gross of 'Booming Ben', the last of his
kind.

Wake Island Rail
Gallirallus wakensis

Wake Island is a tiny spot of land – V-shaped and only about 3 km (2 miles) in length – that lies way out in the Pacific Ocean, approximately two thirds of the way between North America and Asia. It is part of a small atoll of three islands (the others are Peale and Wilkes Islands) that nowhere rises more than 6 metres (20 feet) above sea level. Like some other inconspicuous islands of the Pacific they became part of a horrifying theatre of conflict during World War II. And this had terrible consequences for one of Wake's avian inhabitants.

Just a few years earlier this small creature – one of the group of birds known as rails – was watched with curiosity by a group of American servicemen, who were stationed on the island to build an airstrip. They would watch with intense interest as the birds attacked hermit crabs. One of them, William Stephen Grooch (1890–1939), wrote:

> *The chief enemy of the hermit crab is a small*
> *bird known as the wingless rail, dusty brown*
> *in colour, a bit larger than a sparrow…The rail*
> *is a meek-looking little chap but a doughty fighter.*
> *I saw one attack a crab with an approach so rapid*
> *that the crab did not have time to clew up in his*
> *shell. The rail led to the crab's chin like lightning*
> *several times and it was all over for the crab.*
> *Whereupon the rail and his friends proceeded*
> *to eat. While … on Peale* [the species was also
> present on this tiny island although not it seems
> on Wilkes, the third island in the archipelago]
> *we pumped five or six hundred gallons of*

water … onto the ground where it formed a large shallow puddle. Dozens of rails took advantage of this opportunity to have a bath. They splashed about and appeared to enjoy themselves hugely.

Grooch (or maybe one of his companions) took a photo, and this was published in his book *Skyway to Asia* (1936). It shows an individual attacking a hermit crab in very much the manner he described.

Many members of the rail family have a great tendency to disperse, and individuals fly enormous distances searching for

(*Above*). A Wake Island Rail attacking a hermit crab. This photograph from Grooch's book was taken by an American serviceman building an airstrip on Wake during the mid-1930s.

new lands to colonise. Members of one species, the Banded Rail (*Gallirallus philippensis*) – or perhaps its ancestors – have flown way out across the Pacific and formed colonies in many far-flung and isolated islands. Once there, these birds adapt to their new surroundings and evolve steadily. Where pickings are easy they develop features that help them to survive more comfortably, and also lose those they no longer require. Often this means that they lose the power of flight, for there may be no mammalian enemies and therefore nothing much to flee from. Eventually these immigrants develop into entities significantly different from their colonising ancestors. They pass the point where they can be identified as new species, even though their original affinities can still be recognised. And this is what happened to the Wake Island Rail. Its wings atrophied (which is why Grooch described it as 'wingless') and it became smaller than its ancestors. It also became paler in colouring and lost the rather spectacular spotting of the ancestral species.

So why should a species specially adapted to its island life become extinct? The answer is simple. It is because times and situations change. With island creatures such change often has to do with the coming of humans and the rapacious predators (rats, cats and especially dogs) they bring with them, either accidentally or by design. In this case, however, the American servicemen that came to uninhabited Wake during early decades of the 20th century seemed to rather like their new companions. But then another factor reared its head – war.

Very soon after the United States entered World War II in 1941 (at the end of 1941) the American base on Wake was overrun by Japanese forces. But in due course the resulting garrison became entirely isolated from the rest of the Japanese military machine, and the soldiers left to fend for themselves.

With no supplies coming in they had to find whatever food they could, and on tiny Wake there wasn't much. Starving men turned their attention towards the rails. With little experience of predatory humans, and rudimentary flight feathers incapable of bearing them to safety, the small birds stood little chance. Although they could probably scuttle away quickly enough when chased by a single person, if two or three men joined the pursuit, the rails were likely to be quickly cornered. One after another they would have been hunted down and eaten. Intense bombardment of the island would also have taken a heavy toll. When US forces re-took Wake near the end of 1945, not a single individual could be found.

(*Above*). Two Wake Island Rails in happier times, enjoying the company of American visitors to their island. A photograph taken by an unknown hand in the 1930s.

Laysan Rail
Porzana palmeri

The story of the Laysan Rail is curious and eventful. The island it once lived on is so remote (the most far-flung of the Hawaiian islands) and so small that there was little reason for mariners to visit, and they did so only by chance. In 1828 a Russian vessel landed and the tiny resident rails were noticed, but they were then left in peace for more than 60 years.

During the 1890s, a peculiar set of circumstances led to an expedition backed by the obsessed natural history collector Walter Rothschild (1868–1937) arriving at Laysan. He wished to produce a book on the birds of the Hawaiian Islands, but two other writers, S. Wilson and A. Evans, were beating him to the punch with a volume they were titling *Aves Hawaiienses* (1890–1899). An intense rivalry developed and Rothschild, in a fit of eccentric pique, felt the need to give his own sumptuous work (full of beautiful hand-coloured plates by the celebrated artist John Gerrard Keulemans) a distinctly different, but rather misleading, name. He chose to call it *The Avifauna of Laysan and the Neighbouring Islands* (1893–1900), these 'neighbours' being the whole of the Hawaiian chain. So it was that this massive work came to be named after a tiny island just two miles long by one mile wide (3 km by 1.5 km) while the much larger and infinitely more important Hawaiian Islands (the actual subject of most of the book) were relegated in the title to the status of 'neighbouring.'

(*Facing page*). The best known photograph of a Laysan Rail. It was taken in December 1912 by Alfred M. Bailey. By permission of Denver Museum of Nature and Science.

(*Above*). A Laysan Rail on its nest. This photograph was taken by Walter K. Fisher in May 1902. By permission of Denver Museum of Nature and Science.

To attempt to justify his selection of Laysan for his title the fantastically wealthy Rothschild had sent his agents and collectors there and instructed them to make an intensive survey. And there, of course, they found the rails. There were lots of them; the population was estimated at around 2,000 individuals. Rothschild's collectors came and went and, apart from taking a few specimens for Rothschild's museum, did no real harm to the population as a whole.

Another ten years or so went by, and the rails remained comparatively unmolested on their island sanctuary. At some point they were visited by an American naturalist named Walter K. Fisher (1878–1953), who took several photos.

Then something extraordinary happened. Someone decided that Laysan was a suitable place for a business venture. The idea was to release rabbits and guinea pigs, farm them, and establish a meat cannery in the forlorn expectation that such an enterprise on this remote scrap of land would represent a commercially viable proposition. Any rational mind might have supposed this to be doomed from the outset, but it went ahead nonetheless. Quite how the logistics were assessed is not known, but the idea failed. So did the island's vegetation.

By 1912 it was apparent that escalating numbers of rabbits were destroying the habitat. In 1923 a Smithsonian-sponsored expedition aboard the USS *Tanager* came to Laysan. During this visit just two rails were seen. One of them is shown in a very short but remarkable film (that today can be accessed on YouTube) running across a patch of sand. It was taken by Donald R. Dickey (1887–1932), the expedition's cameraman.

By 1925 the island was utterly ruined. In an article for the *National Geographic,* Alexander Wetmore (1886–1978), an ornithologist who eventually became Secretary of the Smithsonian Institution, described a visit to Laysan:

> *On every hand extended a barren waste of sand … The desolateness of the scene was so depressing that unconsciously we talked in undertones. From all appearances, Laysan might have been some desert.*

Soon the rails were gone.

But this was not quite the end of the story, for in 1891 a few birds had been taken from Laysan and re-located on Eastern Island in the Midway Atoll, some 480 kilometres (300 miles) away. During 1912 Alfred M. Bailey (1894–1978), fearing for Laysan's future, visited the island and took a few more to Eastern. Bailey was a character who interested himself deeply in the plight of endangered animals and participated in a number of expeditions to discover more about them. For many years he worked at the Denver Museum of Nature and Science and produced a series of books under the auspices of that museum, charting his work. Keenly interested in photography, it was Bailey who was responsible for many photographs of endangered birds, and a few of populations that are now extinct.

His efforts to secure the future of the Laysan Rail were, however, destined for failure. Although the birds on Eastern Island flourished, they too became casualties of World War II like their Wake Island counterparts (see page 46). During 1943, a US Navy landing craft accidentally drifted ashore at Eastern bringing with it an invasion of rats. By the end of the war all trace of the rails was gone.

Meanwhile, the rabbits on Laysan itself had been exterminated and the vegetation was almost fully restored. But by now there were no rails left to take back.

(*Facing page*). The nest and eggs of an extinct bird. Laysan Rail eggs photographed by Walter K. Fisher in May 1902. By permisssion of Denver Museum of Nature and Science.

Eskimo Curlew
Numenius borealis

The Eskimo Curlew is one of those species that occurred in vast numbers during the first half of the 19th century but whose population somehow crashed in dramatic fashion. It quickly became excessively rare, and now seems almost certainly extinct. There are those who believe the species still exists though, and there is some hope that this might actually be so. Eskimo Curlews look very much like some of their near relatives, most notably the Whimbrel (*Numenius phaeopus*) and the Little Curlew (*Numenius minutus*). Anyone seeing a living Eskimo Curlew in the field might easily suppose that he or she was looking at an individual of one of these other species. So it is possible that birds could be overlooked. However, it is true to say that most people who study wading birds believe that the species is gone.

One of the last places from which it was reliably reported was Galveston Island, Texas, and it was there that some photos were taken. In April 1962 Don Bleitz (1915–1986), a man who devoted much of his life to photographing birds, managed to take the only known photos of living Eskimo Curlews. However, questions have been asked about the validity of these images, and there are suggestions that they are not all they seem to be. There are several pictures, and some have suggested they are photos of stuffed birds. In fairness to Bleitz it can be said that they do not look like images of preserved specimens, and that they date from a time long before the advent of Photoshop and easy image manipulation. The pictures are sometimes reproduced in colour; whether this

results from enhancement of original black-and-white images is difficult to determine. Bleitz did write fairly convincingly about the circumstances surrounding the taking of his photos:

> *We could almost always find the bird within a few minutes of arriving. I approached to within 40 feet … I was photographing a male when I suddenly became aware of another standing almost beside the one I was photographing. Immediately the first bird, which had been using this feeding territory, ran at the new arrival and continued to jab at him with its bill until he was able to flush the new individual.*

(*Above*). One of the photos taken by Don Bleitz during 1962 at Galveston Island, Texas. Recently, these photos have caused a certain amount of (perhaps unfair) controversy with some commentators suggesting that they might be photographs of stuffed birds.

(*Above*). Three images taken by Don Bleitz that have recently come under suspicion. The image at top left may be the same as that on the previous page, reversed and slightly manipulated. The larger image and that at top right are said by some commentators to be a stuffed bird taken from different angles. The truth is now impossible to determine, but there is no entirely convincing reason to doubt Mr. Bleitz's honesty.

In its heyday, this was a species that performed a remarkable migration, undertaking an epic journey twice-yearly. During late summer, birds would leave their breeding grounds on the north-western corner of Canada and coasts of Alaska, and cross North America to Labrador and Newfoundland. Then they would head out over the Atlantic and make for South America to winter in Argentina. In February or March they would start the return journey but, curiously, they took an entirely different route. They made for Texas and Louisiana, then proceeded north along the Mississippi, Missouri and Platte valleys. Late in May they began to reappear at their breeding grounds.

The reason for the species' headlong rush to extinction is something of a mystery. After all, the Little Curlew leads a similar lifestyle, migrating annually from Siberia to Australia and back again, and yet it still survives in numbers. At least part of the cause for the decline of the Eskimo Curlew must have been excessive hunting. During their flights of mass migration these birds made an easy target for men with guns. The flocks consisted of huge numbers of birds and the route that they would take was well known, as was the likely time when they would occur in a given area. An illustrious 19th century ornithologist named Elliott Coues (1842–1899) wrote a book called *Birds of the North West* (1874). In it he outlined the sort of thing that happened:

> *They generally flew in so loose and straggling*
> *a manner that it is rare to kill more than half a*
> *dozen at a shot. When they wheel, however,*
> *in any of their more beautiful evolutions, they*
> *close together in a more compact body and*
> *offer a more favourable opportunity to the*
> *gunner.*

Mass killings were a commercial operation, and in many aspects the slaughter of so many birds paralleled the hunting of Passenger Pigeons (see page 62).

Yet the species never achieved the celebrity of the other North American species that suffered calamitous declines through hunting. Perhaps this is because a few individuals lingered on for so long; perhaps it is because these birds look so similar to some of their close relatives. What fame the species has achieved is largely due to a work of fiction. A Canadian journalist named Fred Bodsworth (1918–2012) wrote an influential novel called *Last of the Curlews* (1954) which charts the quest of a male Eskimo Curlew looking for a mate. The novel spawned several imitators featuring other extinct and endangered birds, and changed Bosworth's life. It sold more than three million copies.

It was around 1870 that the precipitous drop in numbers started to become apparent, and the fall was rapid. By 1900 the species was all but extinct. It is a curious and quite inexplicable fact that small numbers of birds somehow clung on and continued to reproduce and migrate for much of the 20th century.

Whether or not there are any individuals left remains a moot point. Nigel Collar, who has been an influential figure at BirdLife International for many years, produced a book called *Threatened Birds of the Americas* (1992). In it he and his co-authors expressed this opinion:

> *The fact that its population never recovered, once it became so small that hunting could not be expected to affect it, strongly suggests that a major ecological factor has been in play, rendering all plans and hopes for the species ultimately in vain.*

(*Above*). A hand-coloured version of one of Don Bleitz's photographs from Galveston Island (courtesy of the Western Foundation for Vertebrate Zoology).

Passenger Pigeon
Ectopistes migratorius

The story of the Passenger Pigeon is such a startling one that it is often told, and in all the annals of extinction there is no other tale quite like it. At the start of the 19th century this fast-flighted, streamlined pigeon was quite possibly the most numerous bird on earth; it existed in staggeringly vast numbers. By the century's end, the colossal population had dwindled almost to zero, and the last record of a wild bird is of one shot by an adolescent boy in Pike County, Ohio in March 1900. A few individuals are known to have survived for a little longer, however, living in captivity in Milwaukee, Chicago and Cincinnati.

But fourteen years into the new century the last of these captive birds died in its cage, and a species that a mere hundred years earlier could be counted in the billions, was extinct.

What could have caused this amazingly rapid and spectacular decline? The plain answer is that we don't really know – at least, not entirely. The main ingredient in the story was certainly over-hunting, but some elements in the tragedy remain mysterious.

It was once estimated that birds of this species made up 40% of the bird population of the United States. In other words, four out of ten of the birds living between Mexico and Canada were Passenger Pigeons. Whether this estimate was anywhere near accurate cannot be said, but the fact that it was made gives some indication of how the sheer numbers impressed all who encountered these creatures.

Early written accounts that describe the arrival of flocks of

migrating Passenger Pigeons are almost unbelievable. This one, for instance, was written in the 1830s by the famous bird painter and author John James Audubon (1785–1851):

> *Suddenly there burst forth a general cry of 'Here they come!' The noise which they made, though yet distant, reminded me of a hard gale at sea … The pigeons, arriving by thousands, alighted everywhere, one above another, until solid masses as large as hogsheads, were formed on the branches … Here and*

(*Above*). Martha, last of the Passenger Pigeons, in her cage at the Cincinnatti Zoo, photographed by either Enno Meyer or William C. Herman.

there the perches gave way under the weight with
a crash, and falling to the ground, destroyed
hundreds of the birds beneath ... I found it quite
useless to speak, or even to shout, to those persons ...
nearest to me. The air was literally filled with
pigeons, the light of noonday was obscured as by
an eclipse; the dung fell in spots, not unlike melting
flakes of snow ... pigeons were still passing in
un-diminished numbers and continued to do so
for three days in succession.

Passenger Pigeons (as their name suggests) seem to have been continually on the move, and whenever flocks arrived in an area vegetation was quickly devastated – after which the birds moved on again. It could hardly have been otherwise with so many individuals needing food. In some respects these birds were the avian equivalent to locusts.

Because of this itinerant lifestyle, the species was both a pest and a target for those settling the mid-west. Whenever the flocks came (and their arrival was entirely unpredictable) they brought devastation. But they also brought an opportunity for a bizarre and brutal form of sport. Passenger Pigeon shooting was like no other kind of hunt. When the vast flocks flew overhead, killing birds was simple. All that was necessary was to point a weapon skywards, fire and then re-load as quickly as possible. In this way thousands could be killed in a single day. Local shooting competitions were organised in which upwards of 20,000 kills needed to be recorded to claim a prize.

Activities such as these were frequent throughout the first half of the 19th century. Small wonder, then, that the population began to plummet, and continued to do so. But you cannot actually shoot a species to extinction in this way. The logistics don't work.

First, over a period of time the birds become scarcer. Then, numbers shot become smaller as the huge multitudes are thinned. Eventually, even though there may still be thousands and thousands of birds in existence, there aren't enough individuals in one place to make the organising of a shooting competition worthwhile.

Finally, shooting Passenger Pigeons became no easier than shooting any other species of bird. In a land as vast as the United States there can be no mopping-up hunting operation for a species as small as a pigeon. An additional factor must therefore have been at work. What this was is not well understood. Various ideas have been put forward – forest clearance and imported avian disease being among the more sensible, with mass drownings and the curse of a Christian minister among the more fanciful. Perhaps the best suggestion is that this species had evolved in such a way that it could only survive in vast flocks. Once its numbers fell below a certain level, even though that level was still unimaginably high, the species was doomed to a downward spiral that led to an extinction that nothing could stop.

What can be said with certainty is that any drop in the population seemed imperceptible throughout the first half of the 19th century, but the decline reached a critical phase during the 1870s. At the start of this decade, huge flocks still existed. By its end the remnants had scattered and the species' hold on survival had been broken forever.

By the century's last decade there were very few pigeons left. The fourteen-year-old boy who killed what may have been the last wild bird was called Press Clay Southworth (1886–1979); he did so on his family's farm in Ohio. The family appear to have realised that there was something special about this bird – a female – and they had it stuffed.

(*Facing page*). A remarkable series of photographs exist showing Passenger Pigeons in the aviaries of C. O. Whitman (1842–1910), once Professor of Zoology at the University of Chicago. These photos were taken during 1896 and are now the property of the Historical Society of Wisconsin. There is some doubt over who actually took them. It may have been Whitman himself, but it is more likely they were taken by someone named J. G. Hubbard. They were somehow acquired by the celebrated ornithologist Frank M. Chapman (1864–1945), who for many years was Curator of Birds at The American Museum of Natural History. At some point Chapman passed them on to the Wisconsin Historical Society.

This particular image shows a chick, or squab as the young of pigeons are sometimes called. There seems to be no record as to whether this individual survived to adulthood.

More images from the series are reproduced overleaf.

The work was rather crudely done and buttons were used for the eyes. Appreciating that this specimen represented something of a milestone – albeit a terrible one – the family gave it to a local museum a few years later. It still exists today, and can be seen at the museum of the Ohio History Center in Columbus.

The death of this bird didn't quite represent the final extinction of the species. A few individuals had been caught several years previously, and these were divided between two pigeon fanciers in Milwaukee and Chicago, with another group in the Cincinnati Zoo. By early 1909 all the Milwaukee and Chicago birds were gone, leaving just three individuals in Cincinnati. These were a male and a female (named George and Martha, after George Washington and his wife), and an additional male. By the end of the summer of 1910 only the now-famous Martha remained.

Several pictures of Martha exist, including the photograph reproduced at the beginning of this chapter, which shows her in her outdoor cage at the zoo. As with many photos of extinct animals the precise identity of the photographer is unknown. It was either taken by Enno Meyer or by William C. Herman. Although its date is unknown, this is probably the last photo taken of Martha while she was alive (many have been taken since her death).

(*Facing page*). Four images of Passenger Pigeons from the series taken in 1896 in the aviaries of C. O. Whitman, once Professor of Zoology at the University of Chicago.

Martha was twenty-five years old at the time of the death of her last companions, and clearly the end for her was unlikely to be long in coming. Yet she lingered on alone for a few more years. Whether this lonely existence was an ordeal for a bird as highly social as a Passenger Pigeon cannot be said.

During August of 1914 it became obvious that Martha's long life was at last nearing its end. On the 18th of that month an article in the *Cincinnati Enquirer* stated:

> [She] *has lived for almost 30 years … under the tender care of General Manager Sol A. Stephan, but he has abandoned hopes of keeping* [her] *alive more than a few weeks longer at the very best. That it has been failing rapidly has been noted for some time, but it was not considered more than the feebleness of extreme old age until yesterday morning when Superintendent Stephan discovered it early in the morning lying on its back apparently dead. A few small grains of sand tossed upon it shocked it into activity again, and last night it was acting stronger and fed heartily when the evening feed was offered.*

There are conflicting reports about the exact time of the final event, but it seems that at about 1 o'clock in the afternoon of Tuesday September 1st 1914, Martha was found lying dead on the floor of her cage. This time there was to be no successful resuscitation; Martha's body was frozen in a large block of ice and she was shipped to the Smithsonian Institution to be stuffed. Feathers that she had recently shed in moulting had been collected, and these were sent along with

the body. In the words of the *Cincinnati Enquirer*, this was so that she could:

> *… be shown to posterity, not as an old*
> *bird with most of her plumage gone, as she*
> *is now, but as the queenly young passenger*
> *pigeon that delighted thousands of bird*
> *and nature lovers at the Zoo during the*
> *past 30 years.*

(*Above*). The most famous of all images of Martha. Despite its rather poor quality this well-known photograph has been reproduced in many books and magazine articles. Nothing seems to be known about the circumstances of its taking.

Carolina Parakeet
Conuropsis carolinensis

There are many ways in which the story of the Carolina Parakeet runs parallel to that of the Passenger Pigeon. These birds were both inhabitants of North America, and were species that needed to live in great colonies. They were regarded as serious pests and slaughtered in vast numbers, a persecution that was waged through the 19th century. At the start of that century both species included millions of individuals; by its end both were virtually gone. And curiously the very last representatives of their respective kinds died at the same zoo in Cincinnati.

The last of the parakeets managed to outlive the last of the pigeons by just a few years – three and a half to be precise. Unlike the last pigeon, this was a male and he had been at the zoo for a long time. At an unspecified date during the 1880s, a consignment of 16 small, green parakeets with bright yellow heads had arrived in Cincinnati, having been purchased for just $2.50 per head. Among this consignment were two individuals destined to become the last two Carolina Parakeets. At the time of purchase several other zoos and aviaries had members of the species, but as the years passed so too did the birds in the other collections, leaving just the individuals in Cincinnati. But these too gradually died off until just a pair were left – a male and a female. Their names were Incas and Lady Jane and they had been cage-mates since their arrival. Realising their unique status, London Zoo had tried to buy them, offering $400 for the pair, but the offer was rejected. Then in the summer of 1917 Lady Jane died, leaving Incas entirely alone. He lingered on for a few more months until the evening of Thursday, February 21st 1918 (despite this seeming precision the date may have been exactly one week earlier – the record is not entirely clear), when he died in his cage surrounded by his keepers. They were clear in their minds about the cause of death. He had died of grief.

Of course, no-one knows for sure that he was actually the very last of his kind. Although it is probable that all wild birds were gone by 1918, there may have been a few still lurking in some wild place, doomed to continue a dwindling existence for a few more years. Indeed, there are a number of alleged sightings on record of parakeets in the wild long after 1918. However, none of these has ever been fully substantiated.

This was the only species of parrot indigenous to the United States, and it ranged across the eastern half of the country from the Gulf of Mexico in the south to New York and the southern fringes of the Great Lakes in the north. This northerly range with its penetration into colder climates was extraordinary for a parrot species, but the ability of parakeets to adapt to harsher climes has been shown in the last few decades by the colonies of escaped birds (for instance, Rose-ringed Parakeets *Psittacula krameri*) that have survived – and seem to thrive – in parts of Europe and North America.

As far as Carolina Parakeets were concerned, however, the one thing they didn't seem able to cope with was the encroachment of humans and the changes they brought to the landscape. As European influence increased as the 18th and 19th centuries wore on, the range of the Carolina Parakeet steadily contracted west towards the Mississippi River and south towards Florida. Forest destruction and land clearance along with excessive hunting seem to have been responsible for this decline. Like the Passenger Pigeon, this may have been one of those species that could only maintain itself in large numbers, and once those numbers had dropped below certain levels a spiral of decline was unleashed that was unstoppable.

Although these parakeets originally fed on the seeds of various wild plants, once people began to cultivate the land in an intensive fashion, attention was naturally drawn to these new food sources. In addition these parakeets seem to have been peculiarly unwary birds that could be approached with ease at their feeding grounds. The result of this was that they quickly became regarded as pests and, accordingly, were hunted without mercy.

There is possibly an interesting additional cause for the species' extinction. Honeybees came to the shores of America with European colonists, and it seems that they favoured the hollow trees that parakeets used for roosting and nesting. The birds may have been simply forced out.

A number of photos of an individual named Doodles are said to exist, but only one has been located. It is often reproduced, but the story of this photograph is less well known. During 1896, a well known ornithologist named Robert Ridgway (1850–1929) returned from a collecting trip to Florida with some living parakeets. Two of them produced young, but one of the chicks was being neglected by the parents. Ridgway gave the poorly youngster to a man named Paul Bartsch (1871–1960) who took it home and cared for it. The chick became known as Doodles, and he became very tame. During 1906, Bartsch apparently took several photos of Doodles, but only one seems available. Eventually, in 1914, this much-loved pet bird expired, dropping from his favourite perch above a doorway. He was picked up and slowly passed away in the gentle hands of his owner. Bartsch wrote of Doodles:

> He shared our meals, was well behaved, and
> stuck to his own plate almost always.

Before Doodles had his picture taken, a Dr R. W. Shufeldt (1850–1934) had managed to photograph another captive. This took place around the year 1900. Shufeldt spent several

(*Facing page*). Doodles, photographed during 1906 by his owner Paul Bartsch. Shown here with a Mr Bryan (about whom nothing is known – but who was, presumably, a family friend), this picture shows just how tame and friendly Doodles was.

hours trying to capture an image of two individuals at his home but in the end he got a picture of just one of them. His photo shows a bird in a tangle of cocklebur and has the look of a picture of stuffed bird, but apparently it is the actual image of a living creature. Soon afterwards, Dr Shufeldt made a bad mistake. He painted the cage that the birds used and they both died after chewing the wires; presumably the paint was laced with lead or some other toxin.

(*Above*). Dr Shufeldt's photograph of one of his captive birds, taken around 1900.

It is curious that only these two photographs of living individuals have been located. Many parakeets were in captivity during the last decades of the 19th century and early years of the 20th, and the unique quality of the last pair was widely recognised. It seems extraordinary, therefore, that no other photographs have come to light. Surely there are more out there somewhere.

There is a further mystery with regard to 'Incas'. When the last Passenger Pigeon – 'Martha' – died, her corpse was frozen in a block of ice and sent from Cincinnati to the Smithsonian Institution in Washington so that it could be properly preserved. The same procedure was followed for Incas, the parcel was sent, but the body never arrived at its intended destination. No-one knows what happened to it.

Paradise Parrot
Psephotus pulcherrimus

During the month of June in the year 1844, an English naturalist named John Gilbert (1812–1845) was exploring the Darling Downs in southern Queensland, Australia. His mission was to collect bird specimens for John Gould (1804–1881), the celebrated producer of many sumptuous bird books. Gould needed material for a work he had in hand, a series of seven huge volumes that was to be titled *The Birds of Australia* (1840–1848). This set of books would eventually contain almost 600 hand-coloured lithograph plates depicting all the species then known to inhabit Australia. It is now regarded as one of the great treasures of ornithological literature.

While Gilbert was enduring all kinds of privations in the wilds of Australia to provide Gould with the reference material he required, Gould was working equally hard back in England, commissioning artists, printers and lithographers, writing text, and drumming up subscribers for his very expensive project.

On June 8th, Gilbert wrote to Gould with some exciting news. He had discovered a small parrot of an entirely new kind. Not only was it new, it was also incredibly beautiful

(*Facing page*). Three images of the Paradise Parrot taken in fairly quick succession by C. H. H. Jerrard during 1922 at Burnett River Queensland. They show the bird perched on a termite mound by its nesting hole. The top image is reproduced from a hand-coloured lantern slide that was made from one of Jerrard's original black-and-white photographs. The poor colouring is a weak attempt to show the beautiful colours of the living bird.

with a remarkable mixture of colours – blue, green, yellow, red and brown. Gilbert had already collected many bird specimens for Gould, but this one inspired him to ask a particular favour. He wondered if the beautiful little parrot could be called *gilberti* when Gould got round to giving it a scientific name. His letter from the outback of Australia took months to reach London, of course, and when it did Gould was delighted to hear of the new find, but for unknown reasons he rejected Gilbert's proposal. He gave the species the scientific name of *Psephotus pulcherrimus* (loosely translated, *pulcherrimus* means 'very lovely') instead of the *Psephotus gilberti* that Gilbert had hoped for. Gould then wrote back to Gilbert explaining what he had done.

Gilbert never received the letter. He was dead. Communication between the two men over the vast distances involved had taken so long that Gilbert had set off on another expedition in the mean time. On June 25th, 1845, little more than a year after his discovery, he was speared during an attack by Aborigines and killed. History doesn't relate whether Gould had any remorse over his failure to name the parrot according to Gilbert's wishes. Given the circumstances he probably did, but we shall never know.

This is how the extinct species now known as the Paradise Parrot was discovered and named. But is it really extinct? This is one of those species for which rumours of survival abound, and there are many who believe that somewhere in the vastness of outback Australia it might still exist.

(*Facing page*). Two photographs taken by C. H. H. Jerrard on April 24th 1922, showing eggs in an abandoned nest in a termite mound. Jerrard opened up the cavity to take the photographs once he was sure that the nest had been abandoned.

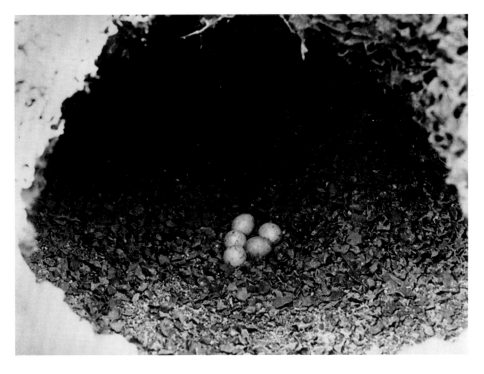

The reality, however, is that it almost certainly doesn't. Claims that individuals survive in captivity have all proved fraudulent, and the areas that the parrot once inhabited have been searched repeatedly. Sadly, there are compelling reasons for supposing that it has vanished forever. Following the initial discovery, birds didn't prove hard to find; they were perhaps never more than locally common, but seemed to be quite widely distributed and occurred in all kinds of suitable areas. A pattern of decline was quickly established, however, and this was regularly commented on over several decades around the turn of the 19th century. Introduced predators such as cats, rats and foxes may have had a detrimental effect, but perhaps even more telling was the use of land for the grazing of cattle and sheep. The seeds that the birds favoured would have become in short supply.

Whatever the reasons, there was a rapid decline in the parrot population during the years leading up to World War I, and by the time war broke out Paradise Parrots seemed to be entirely gone.

This was an inhabitant of valleys, savanna woodland and scrub grassland in southern Queensland and northern New South Wales. Nests were sometimes hollowed out in steep or vertical river banks, although more usually the birds made tunnels in termite mounds. But in many apparently favourable nesting sites there was no longer any sign of them. As the war ended a well-known Australian ornithologist, A. H. Chisholm (1890–1977), launched a newspaper campaign aimed at attracting attention and hopefully locating any surviving population.

(*Facing page*). The first photograph (and probably the best) of a Paradise Parrot taken by C. H. H. Jerrard at Burnett River, Queensland. It was taken on March 7th 1922.

There was no response. Not at first, at any rate. Then, three years after he had begun his campaign, Chisholm received a letter from a Mr Cyril H. H. Jerrard (1889–1943). Mr Jerrard had seen Paradise Parrots! And not only had he seen them, but he had hatched a plan to photograph them, too. Fortunately, he left a very detailed account explaining just how he went about things:

> On March 7th [1922] I pitched my camp before the little clay fortress [the birds' nesting site] … The tent was a small cubical affair about 3 ft 6 ins [just over a metre] in each of its three dimensions and made of old weather beaten bagging. I had sewn it together at home so that its erection entailed only the cutting of four short stakes … and driving them into the ground at positions corresponding to the corners of the tent … A small aperture at the front formed a loophole through which the photographic 'gun' was 'aimed.' These preparations for bloodless 'shooting' were made about noon. Then the new ornament to the landscape was left for a couple of hours in order that the parrots might familiarize themselves with it while I adjourned for lunch. When I came back … I went into hiding with a mixture of hope and dubiety. It was a hot afternoon and my place of confinement was small and ill-ventilated … Ere an hour had passed, however, there came a magic sound … the 'qui-vive' note of the male Paradise Parrot. He was in a tree close to me, but I

(*Facing page*). The tent that Jerrard used as a hide to facilitate taking his photos.

*could not see him till … he alighted in all his
glory on the nest mound itself. It was one of the
supreme moments of my life. I pressed the release
and at the slight click he hopped back … But he was
not really alarmed, and I had barely time to change
the plate before he was back on the mound. I
waited. The female had now come into view.
The male approached the nest hole, just where I
wished him to pose, uttered a sweet inviting
chirp to his mate and peered into the hole. In
answer … the female alighted on the summit of the
mound … I 'fired' again, both birds posing for just
the instant required.*

Although he tried again several times (and managed to take more shots of the male), Jerrard was never able to capture an image of the female again. He remarked:

> *Although I attempted a shot, it resulted in a hopelessly blurred image on the negative. This timidity as compared with her mate's boldness … was rather unexpected, because contrary to rule, a female bird, particularly when she is plainer than her mate, usually shows less fear than he about approaching the nest when men are about it … At sundown I packed up my camera, leaving the ungainly tent in position for future use, and went home feeling as triumphant as a general who has won a campaign.*

Jerrard returned to the site on other days and managed to take more pictures. In fact he continued to see Paradise Parrots for several years, but it seems he had his final sighting during 1927. One of his neighbours saw a lone male bird on November 20th 1928, and this may well constitute the last legitimate view of the species.

But the optimists may be right. Other supposedly extinct animals have returned from the 'dead.' Perhaps the Paradise Parrot will join them.

(*Facing page*). Jerrard's photograph taken on March 7th 1922 of the male and female (top) together – the only time he managed to get an image of the female bird.

87

Laughing Owl
Sceloglaux albifacies

At some time during the year 1909, or maybe even a year or two later, two brothers, Cuthbert and Oliver Parr, decided to photograph a Laughing Owl. They knew that one nested in a crevice in a rocky bluff just a short distance from their home at Raincliff Station, South Canterbury, New Zealand.

So, one evening they climbed up to the nest site, just a hole under a limestone boulder, with a quarter-plate camera, a tripod and a dead mouse. On peering into the hole they saw a nestling, almost fully developed. Carefully, they lifted it out and offered it their dead mouse. With some difficulty they persuaded the young bird to grasp it in its beak and when it did they took a photo, then quickly took another. The Parr family kept these photographs for decades, but long after the two brothers had passed away their descendant Dr J. C. Parr made them public. In 1990 he allowed them to be taken to a local newspaper and reproduced. They are the only known photos of the species taken in the wild.

Another photograph does exist, however, but this is of a captive individual. It was taken during 1892 by a Mr Henry Wright, a businessman from Wellington who concerned himself very much with the plight of New Zealand's threatened avifauna and made great efforts to secure land for bird sanctuaries. When Wright heard that two captive Laughing Owls were in the possession of Sir Walter Buller (1838–1906), the celebrated 19th-century chronicler of New Zealand's birds, he got permission to take a photograph. His mission was urgent, for he soon learned that Buller had sold his birds to

Walter Rothschild (1868–1937) the well-known collector of
all things connected with natural history, and that they were
about to be shipped from New Zealand to Britain.

(*Above*). One of the photographs shot by Cuthbert and Oliver Parr.

(*Overleaf*). The better of the Parrs' two photographs.

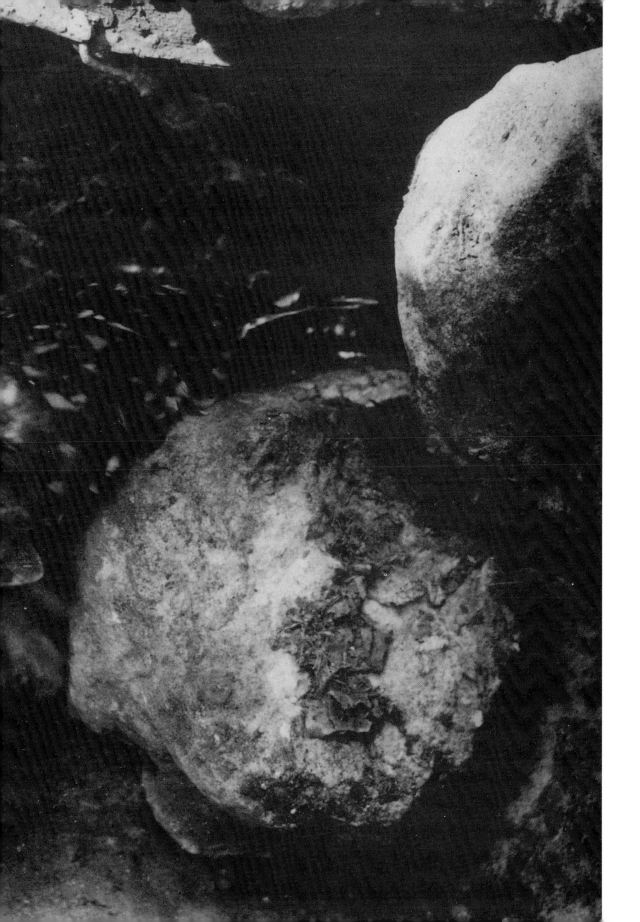

Wright succeeded in his photographic quest but only partially, for he managed to get a good picture of just one of the birds.

A number of other laughing owls were in captivity at one time or another, but this seems to be the only captive individual that was never actually photographed. The ease with which this species (rare even by the mid-19th century) was captured seems to indicate that it was sometimes subject to a rather curious docility. This peculiarity perhaps accounts for the way in which the Parr brothers were able to handle their nestling. In his famous book *A History of the Birds of New Zealand* (1888–89), Walter Buller recorded an instance of the apparent tameness of one particular individual:

> *A man … travelling from Nelson to the West Coast … observed an owl squatting on the ground near the roadside … He dismounted from his horse and caught the bird. Then … he drove a thick pole into the ground and secured his captive by the leg, allowing a sufficient length of flax to permit of the Owl moving freely over the ground. On his return … two days later he found that the bird had snapped, or in some way had got disengaged from the flax string, and was perched on the top of the pole, permitting itself to be recaptured without the slightest resistance. He took it with him to Nelson and, not knowing its value, sold it to the narrator* [Buller] *for a few shillings.*

(*Facing page*). Henry Wright's photograph of a captive individual, taken in 1892. An alternative name for the Laughing Owl was the White-faced Owl, and Wright's picture clearly shows this feature, although it seems that not all individuals had it.

Apparently some individuals had a strange attraction to music. One early European settler in New Zealand maintained that he could always attract Laughing Owls with the sound of his accordion. Soon after he started playing an owl would fly over, perch facing him and remain within hearing until the music had finished.

Although they could be fierce in certain circumstances, it was probably this docility, combined with the fact that they were inclined to spend much of their time on the ground, that made these owls susceptible to attack from introduced ferrets, weasels and cats and, ultimately, hastened their extinction.

For reasons that are not apparent, Laughing Owls seem to have been rare throughout the period of European colonisation of New Zealand, and may even have been in decline long before that. It is perhaps surprising that this owl evolved in a land devoid of small terrestrial mammals, the normal prey of most owl species. Laughing Owls originally fed on a wide range of animals, but mainly on small ground-dwelling birds. Later, the owls were able to feed on rats and mice, too.

The Laughing Owl became steadily rarer throughout the 19th century. Although once present on New Zealand's North Island, by fairly early in European times, they seem to have been largely confined to the South Island where favourite nesting sites were (as the Parr brothers observed) in crevices and fissures on rocky cliffs. The comparatively short wings can be measured in preserved specimens and are an indication that this owl was not a particularly strong flyer. Longish legs and short toes show that it was well suited to moving on the ground.

Most people who hear of the Laughing Owl ask the obvious question – did it actually laugh? The answer is not straightforward. Walter Buller, who obviously handled living individuals, said that it did. According to him it was, 'a peculiar kind of laugh in a descending scale, and very ridiculous to hear.' This was usually uttered when the birds were on the wing, and Buller added that on very dark nights 'when several are hunting together, they seem to laugh in unison.' On the other hand some early writers described hearing nothing more than a series of dismal shrieks that would cause a camper, caught at night in some wild place, to wake with a shudder. This was particularly noticeable – apparently – on dreary nights when it was just starting to drizzle.

The last widely accepted record of a Laughing Owl is of one picked up dead at Blue Cliffs, South Canterbury by a Mrs Airini Woodhouse (1896–1989) in July 1914. Realising the importance of her find, she arranged for the bird to be stuffed and for many years it could be seen at the family home under a large glass dome. It is doubtful if this really was the last one, however. For many years after 1914 people claimed to see individuals. In fact, Oliver Parr recalled seeing birds until about 1924 and, given his photographic credentials, it would be hard to doubt him. Then, in 1938, after being away for several years, he took his son back to the old family home to see if they could find owls.

But they were gone.

Ivory-billed Woodpecker
Campephilus principalis

The likelihood that Ivory-billed Woodpeckers still survive on the North American continent is on the one hand slim, and on the other non-existent. And slim, as they say, just left town. The notion that the species survives flies in the face of common sense and all scientific rigour. Yet in April 2005 newspaper headlines the world over proclaimed that it did, and that a year earlier it had been seen again after going missing for more than half a century. No ifs, no buts, Ivory-billed Woodpeckers still lived in the Arkansas bottomlands! Well, at least one *seemed* to. A bird had been seen by experienced ornithologists (who, apparently, broke down and cried at this emotional moment) and, what's more, it had been conclusively filmed. Or so the headlines announced.

Actually, it hadn't. By any standard of reasonable judgement the photographic evidence was completely unconvincing. For all that any ordinary person could tell, the pictures might have shown a garden gnome riding on the back of the Loch Ness Monster. Most newspapers and magazines elected not to print stills from the video – presumably on the grounds that a sight of the actual evidence might thoroughly spoil a good story.

One of the complications concerning recently extinct species are the alleged sightings often reported after an animal's disappearance. Mostly, these reports centre around

(*Facing page*). A photo taken by James Tanner in April 1935 showing an adult male Ivory-billed Woodpecker returning to a female.

a combination of wishful thinking, genuine mistake and lack of relevant experience or knowledge. Sometimes, they are based on the desire for a strange kind of brief celebrity, and the stories are entirely spurious. Occasionally, of course, there may be real substance to them. And herein lies the problem – sorting the wheat from the chaff.

In the case of the Ivory-billed Woodpecker, there have been a number of alleged sightings (generally dismissed by the ornithological community) since 1944 when the celebrated bird painter Don Eckelberry (1921–2001) made what may well have been the last genuine sighting of an ivory-bill on the North American continent. Why the claimed sighting of 2004 was taken so much more seriously than these other reports (and picked up the following year by newspapers that generally have little or no interest in such matters) is, in itself, a mystery.

The reasons for supposing the rediscovery to be entirely bogus, and for regarding the species as extinct, are many. Prominent among them is the existence of the Pileated Woodpecker (*Dryocopus pileatus*). This is a smaller species than the ivory-bill but its coloration is very similar, and to the uninitiated (and who isn't uninitiated after a 60 year lapse?) the two could be easily confused, particularly if seen fleetingly, or in a situation where scale is difficult to judge.

There is also the fact that ivory-bills left distinctive traces to show their attacks on rotting stumps and trees – and no such traces have been found for decades, despite the United States being home to thousands upon thousands of active birdwatchers, many of whom would be only too delighted to find evidence of the species' continuing existence. Equally important is the distinctive call of the ivory-bill, which carried for a distance and could be heard when there was no chance of actually seeing the individual making the sound.

This was a spectacular black-and-white bird, with males sporting a dramatic red crest at the back of the head. It was large for a woodpecker (around 50 centimetres – 20 inches – long); in fact it was the second-largest of all known woodpeckers, the largest being Mexico's Imperial Woodpecker (see page 112) – which is also now considered extinct.

Ivory-bills were in decline from the point at which European civilisation began making inroads into what is now the south-east United States. Throughout the 19th century numbers dropped and dropped, and they continued to do so during the first decades of the 20th. Although great stands of seemingly suitable forest and swamp habitat still remain even to this day, human interference (or even proximity) appears to have been something that birds of this species could not withstand.

Most of what is known of the living bird comes from the work of James T. Tanner (1914–1991), who located ivory-bills in the 1930s, at a time when the species was all but gone and the population was down to a handful of individuals.

Locating these was not necessarily easy, but neither did it prove exceptionally difficult, and this is another fact that makes the species' survival seem so unlikely; even when desperately rare, the woodpecker proved quite findable. Tanner studied the birds he found for several years, and in 1942 published the results of his intensive study in a substantial monograph.

Before that, on Sunday, March 6th 1938, he temporarily removed a young individual from its nest to ring it, and while it was under his control several wonderful photographs were

(*Next two pages*). Two photos taken by James Tanner on Sunday March 6th 1938 of a young bird taken from its nest for ringing, and playing with Tanner's colleague Joseph Jenkins (Jack) Kuhn. They named the bird 'Sonny Boy'. All pictures of Sonny Boy with J. J. Kuhn are reproduced by kind permission of Nancy Tanner.

taken. He returned the bird to the nesting hole, and twelve days later watched it leave the nest under its own steam, looking very much as it had when he photographed it.

The story of these photos has a remarkable conclusion. Two were reproduced in Tanner's book, *The Ivory-billed Woodpecker,* published in 1942. For more than 60 years most of the ornithological world believed this was all there were. Then a naturalist called Stephen Lyn Bales decided to write a book about Jim Tanner and his exploits. By coincidence he lived in Knoxville, Tennessee, and so did Nancy, Jim's widow. Bales visited her in June 2009, and she showed him the contents of an old brown envelope that she had kept for many years. The envelope contained photographic negatives, that inconvenient relic of the past that younger generations have never needed to worry about. When examined, these negatives held several more images of the young ivory-bill in playful mode with Tanner's colleague J. J. Kuhn, all taken on the same far-away day in 1938. They provide a truly moving record of humanity's last interactions with a dying species.

Mrs Tanner, today in her nineties, had accompanied her husband many years before on some of his field trips. In her foreword to Stephen Bales' book *Ghost Birds* (2010) she described the difficulties that a photographer had to overcome in order to get pictures:

When Jim wanted to take a picture, he had to get close ... because he had no zoom lens. Then he had to use a separate light meter, estimate the distance, set the f-stop and shutter speed, all before he could compose the picture, and most birds did not pose and patiently wait for their portrait!

(*Facing page*). Another of Tanner's photos, this one taken during April 1935 and showing a male returning to the nest to relieve the female.

Despite Tanner's ability to find the birds and the recommendations he made for securing the species' survival, ivory-bills vanished – entirely – very soon after the time of his research.

His main hope for the species centred around a Louisiana forest of some 80,000 acres known as the Singer Tract. This name came about simply because the area was owned by the Singer Sewing Machine Company. With logging activities making serious inroads into the tract, the National Audubon Society fought to keep the forest intact and in 1943 Franklin D. Roosevelt (1882–1945), then President of the United States, was apparently approached to intervene. Perhaps because World War II presented him with more pressing challenges he failed to do so and logging continued. Curiously, German prisoners of war did much of the cutting and a substantial proportion of the wood was used to make tea boxes so that British soldiers could enjoy their favourite drink on the front line.

Very reluctantly, James Tanner came to the belief that the Ivory-billed Woodpecker was gone, but he wrote a rather poetic testimonial to the species:

> [It] *has frequently been described as a dweller of dark and gloomy swamps, has been associated with muck and murk, has been called a melancholy bird, but it is not that at all – the ivory-bill is a dweller of the tree tops and sunshine; it lives in the sun ... in surroundings as bright as its own plumage.*

(*Previous four pages and facing page*). Some of the series of photos takes by James Tanner showing his colleague J. J. Kuhn with the young ivory-bill they named Sonny Boy. All reproduced by kind permission of Nancy Tanner.

The world being full of mystery, there are those who would argue that it is possible – just possible – that the species still survives, but anyone wondering if it does might ask themselves the following series of questions.

Why should a species plummeting to extinction for more than a century suddenly stop its dreadful downward spiral precisely as it reached the edge of oblivion?

Why would the dismal pattern of decline stop during the late 1930s just because numbers had dwindled to no more than a few pairs?

What force is there in nature that would apply a last-minute emergency brake to this headlong rush to disaster and dictate a fierce – yet entirely secretive – rearguard action?

What conspiracy could there be that would motivate these last few individuals to collectively decide that the decline had gone far enough, and that they owed it to their forebears – and the world at large – to secrete themselves away and surreptitiously breed in small and unnoticeable numbers (all the time in silence) and thus perpetuate their race?

Why should these last few pairs, and the pitifully few descendants they may have had, be able to withstand the pressures that decimated their fellows, and then linger on – more or less unnoticed – for upwards of half a century?

What then is left of the Ivory-billed Woodpecker? Apart from stuffed specimens, there are some old photos. There is even a grainy piece of antique nitrose movie film (most of which combusted and was destroyed at Cornell University during the 1960s due to its highly flammable nature). There are a few recordings of the actual call.

And there was also a race of the species that lived in Cuba and was last seen perhaps as recently as the 1990s. But that is another story.

(*Above*). Several days after James Tanner photographed and banded Sonny Boy, he and his colleague saw him again. This was not the last time he was to see the bird. He saw him a year later on another visit to Louisiana. It is not beyond the bounds of possibility that Sonny Boy outlived any others of his kind.

Imperial Woodpecker
Campephilus imperialis

The Ivory-billed Woodpecker (see page 96) had a close relative that lived in the Sierra Madre Occidental of Mexico. Slightly bigger than the ivory-bill, this was the world's largest woodpecker, and so it was called the Imperial. There were many similarities between the two species although, curiously, Imperial Woodpeckers had adapted to high-altitude pine forests whereas ivory-bills inhabited swampy bottomlands. But even though they lived in different environments, both were doomed to become extinct at around the same time. The last confirmed ivory-bill sighting occurred in 1944 while the last believable sighting of an Imperial was made in 1956. A major difference between the two species' history with humans lay in the photographic record. Unlike with the ivory-bill, it was always supposed that there were absolutely no photographs of the Imperial Woodpecker.

Then, in 1997, more than 40 years after the species had vanished, it was revealed that actually there *were* pictures. But these were not just still photographs – this was film footage! And it had been taken during the last fully recognised sighting of the species in 1956.

(*Facing page and above*). Two stills frozen from the film footage obtained by William Rhein in 1956 in the Sierra Madre Occidental, Mexico. After having been overlooked for 40 years or so, the footage was rediscovered and made available in the 1990s by Martjan Lammertink and Cornell University.

As might be expected, the tale of the film's discovery is
a curious one. A Dutch ornithologist, Martjan Lammertink,
had an intense interest in woodpeckers, and was undertaking
research at Cornell University's Department of Ornithology. It
was natural that he should select Cornell, as this was an
institution that had supported James Tanner's book on the
ivory-bill, and it was also the place where many of
Tanner's research records were kept. As he went through this
material, Lammertink came across a letter written to
Tanner in 1962 from a dentist named William Rhein. In the
letter, Rhein mentioned that he had shot some film in Mexico
during 1956 and that it contained:

> *Very poor footage of a female* [Imperial
> Woodpecker] *with several short flight shots
> taken hand held from the back of a mule.*

By the mid-1990s Rhein was in his late 80s, and
Lammertink had a good deal of difficulty in tracing him.
He finally tracked him to his home in the intriguingly named
town of Mechanicsburg, Pennsylvania. The two men were
introduced and eventually sat down together to watch 85
seconds of film, shown from an old 16mm projector.
Naturally, the quality is not good but the film (today
viewable on YouTube) clearly shows a female bird climbing a
pine trunk, foraging, chipping off bark, and then flying away.

Rhein promised to make a copy of the film, but
Lammertink didn't receive it until a couple of years later –
after the Mr Rhein's death. The reason why he had never
made the film available to the ornithological community
before this time remains a mystery. Perhaps its poor quality

embarrassed him; perhaps he just thought that no-one would be interested.

The species itself is almost as poorly known as the film was, although once it was not uncommon. However, the almost total destruction of the open pine forest environment in which it lived brought about its spiral to oblivion. There are certainly instances where hunting has been a major cause of

(*Above*). Another still from Rhein's film, showing the female Imperial Woodpecker flying away.

extinction, but habitat destruction or alteration is the usual culprit.

It is easy enough to point to exceptions (and there are a number in this book), but as a general principle hunting tends to affect individuals rather than whole populations, whereas habitat loss means inevitable destruction for almost all the animals (and plants) that naturally occurred in an area. Only those able to rapidly adapt and take advantage of their new circumstances can survive.

In this particular instance, logging and general land clearance robbed the woodpeckers of the homes and food supplies they needed, and although hunting by indigenous peoples may have accounted for many birds, the species was essentially doomed by the alteration of the landscape. In fact those responsible for changing the environment actively encouraged the hunting of the birds. They saw them as an inconvenience to their operations, and even promoted the application of poisons to the trees in which woodpeckers foraged – trees that were themselves doomed to ultimate destruction.

As part of his ongoing search for any hint of evidence regarding the species' survival, Martjan Lammertink decided to make some on-the-spot inquiries. In 2010, along with his colleague Tim Gallagher (who has written a book, *Imperial Dreams*, on his exploits), he made a long, arduous and dangerous – due to political and criminal activity in the area – journey to the Mexican state of Durango, where the film footage was shot, and the place where the high-altitude pine forests used to grow. The two men collected several stories indicating that the woodpeckers survived long after the 1950s, but nothing tangible to suggest that they might still be there.

As for the pines, they were long gone, and the grasses that once grew beneath them had been grazed away.

(*Facing page*). A last still from Rhein's film.

New Zealand Bush Wren
Xenicus longipes

Don Merton (1939–2011) is a legendary figure in the story of animal conservation and the preservation of species. He was instrumental in saving New Zealand's Black Robin (*Petroica traversi*) from extinction, one of the most spectacular achievements of the entire conservation movement. At one time this species was reduced to just five individuals, three males and two females. Such a small gene pool usually means imminent extinction no matter what efforts are made to enable surviving individuals to reproduce. Somehow Merton and his colleagues managed to overcome the many difficulties involved and today there are more than 200 birds. The Kakapo (*Strigops habroptilus*), the world's largest parrot – and perhaps the most extraordinary – is another species that has clung to existence, and its survival is due – at least in part – to Merton's work. Sadly, neither he nor his colleagues were able to save the New Zealand Bush Wren.

During the early 1960s, however, he took a remarkable photograph, one of two special pictures that he was able to take around this time. One photo was of the very rare Greater Short-tailed Bat (*Mystacina robusta*). The other was of what was then classified as one of the world's most endangered birds, the Bush Wren.

(*Facing page*). Don Merton's photograph of one of the last of the New Zealand Bush Wrens, taken in September 1964 on Big South Cape Island.

At the time he took the photographs, Don Merton was working on one of New Zealand's offshore islands. It is known as Big South Cape, and is a small piece of land off the coast of the much larger Stewart Island. As part of the New Zealand government's attempt to save the nation's endangered species, the island was being maintained as a rat- and predator-free sanctuary, and it was one of the last refuges of both bat and bird. In September 1964 Merton was able to photograph one of the few surviving wrens, but unfortunately

(*Above*). A photograph of Bush Wren eggs taken in 1913 on Big South Cape Island by Herbert Guthrie-Smith, an important figure in the history of New Zealand ornithology.

in that year rats somehow managed to invade the island. It didn't take long for them to do their work. The last fully authenticated records of living wrens in this area date from 1972, and by that time the species seems to have vanished from every other part of New Zealand.

It had once been an inhabitant of much of the country. The Bush Wren was found on both of New Zealand's main islands, although in historical times it seems to always have been rare on the North. On the South Island, however, it was comparatively plentiful. Its small size and a fatal liking for spending time on the ground made it vulnerable to attacks from mammalian predators, of course. Once these animals began to occur on New Zealand soil – through the agency of humans – the Bush Wren's days were numbered. By the beginning of the 20th century it had become rare, and as the century wore on it became excessively so.

Three races were identified, one from the North Island, one from the South, and one from Stewart Island and the small islands near to it. The recorded differences between them were very slight, and with so few museum specimens now available for study it is hard to make meaningful attempts to determine whether or not the alleged differences have any significance.

There are few descriptions of how these birds behaved in life. Perhaps the clearest was penned by the celebrated chronicler of New Zealand's birds Sir Walter Buller in the second edition of his *History of the Birds of New Zealand* (1888–89):

> [They are] *generally met with singly or in pairs
> ... attracting notice by the sprightliness of their
> movements. They run along the boles and
> branches of the trees with restless activity, peering
> into every crevice and searching the bark for the*

*small insects, chrysalids and larvae on which they
feed ... [They have] a weak but lively note, the sexes
always calling to each other with a subdued trill
and ... powers of flight are very limited.*

Don Merton's 1964 picture of one of the last individuals
is not the only photograph in existence. Several were taken on
the same island some 50 years earlier. Another important
figure in New Zealand ornithology, Herbert Guthrie-Smith
(1861–1940), managed to shoot several photos (and also
one of the species' eggs) on a visit he made during 1913, and
he published these in his book *Bird Life on Island and Shore*
(1925).

(*Facing page and above*). Two of Herbert Guthrie-Smith's photographs taken in 1913
and showing a Bush Wren. In the first photo the bird is carrying a feather in its bill,
presumably for nest material.

He also described some of the difficulties he experienced in taking his photos. The island was remote from any of the conveniences of civilisation; the biggest problem was a lack of clean water with which to wash his photographic plates. The birds themselves seemed comparatively undisturbed by the camera and anyone operating it.

Nor were these the only photos to be taken. Several years later Edgar Stead (1881–1949), a third significant figure in the history of New Zealand ornithology, visited nearby Solomon

Island and managed to capture a bird on film. But perhaps the last word on the species should be left to Guthrie-Smith, who wrote:

> *He passes through the darkling underscrub*
> *like a forest gnome, like a woodland brownie.*

(*Facing page*). Two more of Guthrie-Smith's photographs.

(*Above*). Details from the two photos opposite.

(*Overleaf*). Edgar Stead's photo taken on Solomon Island in November 1931.

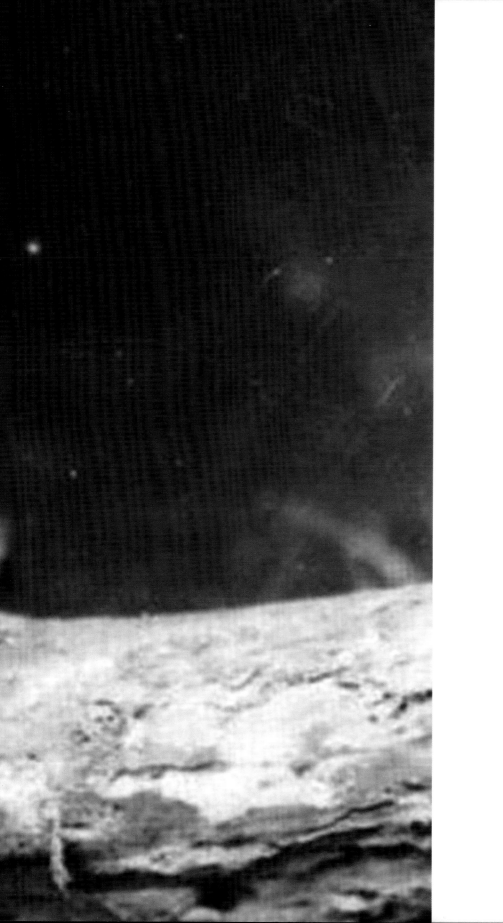

Aldabra Brush Warbler
Nesillas aldabranus

Robert Prŷs-Jones runs the bird department of London's Natural History Museum. Curiously, this particular section of the vast and ramshackle institution isn't situated in London at all, but is located instead thirty miles or so to the

(*Above*). One of Robert Prŷs-Jones's photographs of the Aldabra Brush Warbler. Not only was he one of a handful of humans ever to see the species, he was also the only one to photograph it.

north in a small country town called Tring. There are perfectly good (even interesting) historical reasons for this isolation, but it can be very frustrating to anyone who turns up at the enormous Victorian temple in South Kensington expecting to see birds.

Robert is an enthusiastic man who also happens to have a streak of extraordinary generosity. In addition to this, he is one of just a handful of people who actually saw a living Aldabra Brush Warbler. So it was perfectly natural that I should turn to Robert when conducting research for a previous book, *Extinct Birds* (2000).

'Is there any chance that the species still survives?' This was the question I wanted to ask him. After all, it needn't be featured in a book on extinct birds if it did.

One Tuesday afternoon in 1999, I travelled to Tring to ask the all-important question. We began by talking generally about his Aldabra experience. Now, the idea of spending weeks on end, alone, on a tiny island far, far out in the Indian Ocean is perhaps not to everyone's taste. But that is exactly what Robert did. In fact, if all the time he spent there is added up, it comes to over two years. His description of being dropped at one end of the island and making his way to the other, knowing he wouldn't see another human soul for weeks, filled me with a kind of despair.

"Were you afraid to be so alone?" I asked.

"Not really. I quite liked it," was the answer. Obviously this man is made of sterner stuff than me, I thought. So I turned our talk back to the essential subject.

"Could the brush warbler still survive?" I asked.

"Well, anything is possible, I suppose," Robert answered, "but almost certainly, given the whole situation and the change in the habitat, the species is gone."

"How many times did you see it, Robert?"

"Several. At one time it wasn't that difficult to find. At least it wasn't if you used tape recordings of its call to attract it. Then on later visits I couldn't find any trace of it."

Having taken up enough of Robert's time, I got up to leave and made for the door. But I didn't reach it.

"I've got a photo," he called.

"A photo?"

"Yes, I've got one," he repeated. "Would you like to use it in your book?"

If ever there was a question to which the answer was obvious – this was it! This was a bird that had hardly ever been looked on by humans, that had vanished within a few years of its discovery, that almost nothing was known about – and now I was being told that there was a photograph in existence that the world at large knew nothing about, and that I could use it in my book.

Yet there was one more surprise in store. One of the last people ever to see an Aldabra Brush Warbler rummaged in a drawer and then tossed across his desk that relic of the past, a 35mm transparency. I held it up to the light, rubbed my eyes and looked at it. There it was, a poorly defined little grey-brown blob in the middle of a large expanse of green. Despite its rather unspectacular aspect it was the grail of extinct-bird photograph hunters – a hitherto unknown photo of a virtually unknown species that had lain in the drawer of its owner for 20 years or so.

"You will take care of it, won't you," said Robert. "It's the only copy I have!"

(*Facing page*). Robert Prŷs-Jones's second photograph of an Aldabra Brush Warbler.

Aldabra is a small coral atoll, made up of four main islands, that lies some 425 km (265 miles) north-west of the most northerly point of Madagascar. Politically, it belongs to the Seychelles although it is hundreds of miles south-west of Mahé, the largest of the Seychelles islands. The whole atoll is no more than 46 km (25 miles) long by 16 km (10 miles) wide, and the brush warbler occupied only a tiny part of this: a stretch of land perhaps 50 metres wide by 2.5 km (1.5 miles) long at the western end of Malabar, otherwise known as Middle Island; at least, it was only ever seen or heard in this one very restricted area. Perhaps it once occupied more of the land.

The warbler was discovered entirely by chance, its discoverers choosing to pitch camp right on its doorstep one day late in 1967. Over the next few weeks three birds were caught, two of which were killed and preserved (causing something of a controversy), while the third was released – presumably unharmed. In 1983 an observation of a lone male constitutes the last record of the species. None has been seen or heard since.

In the years between 1967 and 1983 a few observations of living birds were made, and during the 1970s Robert Prŷs-Jones made his studies. He says that at first it proved impossible to find the brush warblers, and only another accidental occurrence enabled it. One of his colleagues, Alex Forbes-Watson, happened to be wandering along a trail in the bush with a tape recorder when he heard a strange bird call. Hearing it again, he decided to press the record button. Curiously, the call was never heard again in this particular area, but it proved to be the song of the Aldabra Brush Warbler and, when played back, it proved invaluable in attracting others of its kind.

Robert doesn't know exactly when he took his photos (there are two) and is unnecessarily hard on himself for failing to keep a proper record. But he does know that it was in 1975.

He recalls that there were considerable difficulties involved in photographing the warblers. It was always difficult to get a sight of the bird in the first place, even when you could hear it. Then, with bright sunlight above and dark shadowy foliage below, any image taken was likely to be unsatisfactory – and unlike today when digital technology makes things so easy and allows instant inspection, there was no way of checking what the camera had captured.

Naturally, a virtually deserted Indian Ocean island lacks photographic development facilities, and it was many weeks, maybe even months, before the film was taken to a laboratory and the results became viewable.

During his stays on Aldabra, Robert went once a month to the spot where the warblers were, and stayed there for a period of four days or so each time. He believes that the species' decline is almost certainly due to the islands having become infested by rats. In the entire period he spent searching for and studying the birds, he was able to locate and definitely distinguish only five individuals. Yet the little that is known of the species is almost exclusively derived from his work.

And, of course, all the photos in existence are his.

Bachman's Warbler
Vermivora bachmanii

(*Above*). Is this an immature female Bachman's Warbler? It was photographed by Robert Barber on March 30th 1977, near to Melbourne, Brevard County, Florida, and may record the last-ever encounter with this species.

Volume 13 of an obscure journal known as *The Florida Field Naturalist* includes articles and papers submitted to the journal's editors during the years 1983–1987. Pages 64 to 66 hold an account that may be much more significant than any of the others, and describe an event that happened several years earlier. The article is a carefully written record of what may have been the last human encounter with a living Bachman's Warbler. Some ornithologists who have researched this species are unaware of the article; others have chosen to disregard it. Yet it is clearly and thoughtfully assembled, and very persuasive. It also contains two photographs.

From 8.15 to 9.30 on the morning of March 30th 1977, an experienced birdwatcher called Robert Barber observed and then photographed a bird that he could not identify, just west of Melbourne, Brevard County, Florida.

It was not until a year later, after much careful examination of museum specimens and inspection of his photos by several seasoned researchers, that Barber concluded he had seen an immature female, and that this female belonged to the species known as Bachman's Warbler. If this was indeed what he saw, then he was probably the last person to ever see one, and so he wrote his account and arranged for it to be published.

The story of Bachman's Warbler begins as enigmatically as it ends. In July 1833 the Reverend John Bachman (1790–1874) found a bird of a kind he had never seen before in a swamp near Charleston, South Carolina. After killing and preserving it, he sent the specimen to a friend who happened to be in the middle of producing what is today the world's most valuable colour plate book. The book is *The Birds of America* (1827–1838), published in four sumptuous and physically huge volumes, and the friend was the famous artist John James Audubon (1785–1851). Upon receiving the bird, Audubon promptly named it after his friend and confidently produced an illustration of it for his book. In this way its image became quite familiar. Yet no similar bird was seen again for more than 50 years.

Then, in 1886, a second individual was shot. A year later the man who had killed this bird shot no less than 31 more examples. Two years passed, and 21 birds of the species flew into a lighthouse. From then on, and for the next few decades, Bachman's Warblers were seen quite regularly, even though it was obvious that they were not common.

(*Above*). The second of Robert Barber's photos.

(*Facing page*). A photograph showing a singing male, allegedly taken on May 15th 1958, near Charleston, South Carolina. Although this picture is often reproduced it has not proved possible to find reliable details concerning it, or another taken just before or just after (see overleaf). Two names have been associated with the pictures, J. H. Dick and Jerry A. Payne, but it is by no means certain who actually took them.

Then, once again, sightings dropped off and by the second half of the 20th century it was apparent that the species was on the verge of extinction. Various reasons for this have been put forward, but the truth is that no-one really knows why.

These were birds that wintered in Cuba and returned to Florida, Georgia, South Carolina and other states in the south-east of the United States for the spring and summer. They usually nested low down among dense bramble growths in forested river bottoms – areas infested with ticks and other small undesirables. In other words, the birds bred in places that are exceptionally unpleasant for people to visit. Maybe this fact gives some small hope that the species may survive. After all, Bachman's Warbler was always comparatively difficult to locate, and did actually vanish for a period of more than 50 years following its initial discovery. The celebrated American wildlife artist Francis Lee Jaques (1887–1969) once said, 'The difference between warblers and no warblers is very small.'

Perhaps therein lies the only hope for Bachman's Warbler.

(*Facing page*). The second of the photographs of a singing Bachman's Warbler said to have been taken near Charleston in 1958.

Kaua´i ´O´o
Moho braccatus

There are two groups of birds on the Hawaiian Islands that are notorious for the number of extinct species they contain. Although these birds are not particularly closely related, they have names that are similar and this sometimes causes confusion. The birds of one group are known as honeycreepers, and the others as honeyeaters. Both names derive from the fact that many species feed on nectar, although most also eat other things like blossom, insects or molluscs.

Birds known by the strange Hawaiian name of ´o´o were all members of the honeyeater family. There were once four distinct, but closely related, species, one on the island of Oahu, one on Molokai, one on Hawaii, and one on Kaua´i; all are now extinct. The species that inhabited Oahu (*Moho apicalis*) was lost early in the 19th century, the Molokai species (*Moho bishopi*) disappeared as the 19th century passed into the 20th, and the Hawaii bird (*Moho nobilis*) vanished during the 1930s. The fourth species managed to survive into more recent times, however. This was the one that inhabited Kaua´i, one of the larger Hawaiian Islands, and it lingered on until the mid-1980s. By the end of that decade it seems that it too was gone.

In appearance, this bird, 20 cm (8 inches) long, was the least spectacular of the four species. The others all sported tufts of yellow feathers on their sides and under-bellies; one even had these on its face. These feathers were among the highly prized plumes that native Hawaiians used by the

thousands in the making of their celebrated feather robes and cloaks, and the Kaua´i species may be considered fortunate to have had only a few of them; they were only present in small yellow tufts on the thighs. Presumably this meant that the bird was not subjected to quite the same hunting pressures as its relatives.

Despite the very small area of ornamental feathers and its less striking appearance, this is the only species that was ever photographed in life, and several pictures of it exist.

(*Above and following two pages*). A Kaua´i ´O´o photographed by Robert Shallenberger in 1975 in the wilderness known as the Alaka´i Swamp just a few years before the species became extinct. Robert is one of the few people alive who ever saw this species. Reproduced courtesy of the photographer.

141

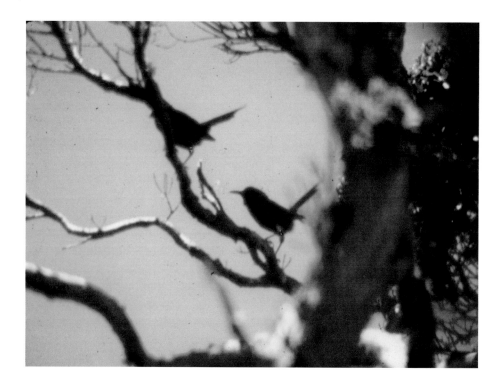

All were taken at a time when the bird was on the verge
of extinction. A dedicated field researcher and photographer
named Robert Shallenberger is responsible for most of them,
and without his efforts there would be no really clear
photographs. There is also a film sequence of an individual
singing (this can be viewed on the internet).

During July 1975 Shallenberger and his colleagues
H. Douglas Pratt and Sheila Conant penetrated a wilderness
known as the Alaka´i Swamp, apparently one of the wettest
places on earth in terms of rainfall. The name 'swamp' is

(*Above*). A pair of Kaua´i ´O´o s photographed during June 1975 by Fred Zeillemaker, a
keen ornithologist from Idaho, during a trip he made to the Alaka´i Swamp. The picture
shows just a hint of the yellow feathers on the species' thigh.

misleading, for the area is not really what is generally understood by that word. It is more of a very wet montane plateau, pocked with bogs and sliced by wooded ravines. In this remote and pristine spot they found the last of the ′o′os. However, it wasn't until 1998 that these long-time students of Hawaiian birdlore published a report on their experiences, and this can be found in a journal called the *Wilson Bulletin* (110, 1–22). In their report they explained the reasons for their delay:

> *At the time we assumed that the relatively large wilderness … would remain a refuge for endangered birds at least for the near future. Thus … we published no general summary of our expedition. Time has shown that our optimism was naïve; we made some of the last observations and tape recordings, and took some of the last (or only) photographs of several species.*

In fact, the end for the ′O′o came very quickly. Just six years after the 1975 visit it seemed only two birds – a pair – remained. Then, in 1983, Hurricane Iwa appears to have accounted for the female, and the male was last seen in 1985. Its haunting, fluting call may have been heard two years later, but after this there was only silence.

At the end of the 19th century, ′O′os had occupied Kaua′i from just above sea-level to the top of Wai′ale′ale, the island's second highest mountain. Within 30 years they were largely gone and only survived in high places like Alaka′i, destroyed elsewhere by avian malaria. This was spread at an alarming rate by mosquitos that were invading the islands. High altitudes soon provided the only mosquito-free areas where the birds could survive, but during hurricanes

or other difficult times the birds were often forced to take shelter at levels the insects could reach. Once bitten, many Hawaiian honeycreepers and honeyeaters would succumb within hours. The few birds that still lived in the high places represented populations that were, ultimately, unsustainable. Following in the wake of dozens of other Hawaiian species that had already gone, the Kauaʹi ʹOʹo disappeared.

(*Facing page*). A photograph of one of the last of the ´o´os. This individual may even have survived for long enough to become the very last. It was taken in 1975 by H. Douglas Pratt, an artist and writer who has produced several important books on Hawaiian birds and who has painted many pictures of them, and accompanied Robert Shallenberger and Sheila Conant on their expedition to the Alaka´i Swamp.

´O´u

Psittirostra psittacea

The Hawaiian honeycreepers provide one of the most clear-cut examples of what is known as 'adaptive radiation.' In the isolation of the Hawaiian Islands, far removed from any major landmasses, the ancestors of this group of birds found that there were many ecological niches left vacant. This was simply because comparatively few kinds of birds managed to reach the islands and colonise them. There were no parrots, for instance, so the hard seeds that many parrots crack with their powerful beaks remained an unexploited food source. So some honeycreepers evolved strong, parrot-like beaks to facilitate crushing such seeds. Others developed bills of quite extraordinary shape. Some were long, slender and down-curved; others developed an upper bill quite different to the lower one. These adaptations enabled the birds to draw nectar or other sustenance from different kinds of flowering plants. So varied did all these adaptations become that it is probably true to say that honeycreepers evolved to fill almost every niche occupied elsewhere in the world by hundreds of different songbird species.

But although birds from this family had adapted perfectly to this multitude of differing niches, the success of each adaptation was dependent on the environment remaining entirely unchanged. Sadly for the honeycreepers it didn't, and the very reasons for their success would spell their ultimate doom.

As humans altered the situation – sometimes in ways that were subtle, other times in ways that were not, sometimes by means that were deliberate, but often by ones that were accidental – species rapidly became extinct, and others rare. Over-hunting, competition from more robust introduced species, introduced avian disease often carried by mosquitos arriving on the islands, and volcanic eruptions or hurricanes that destroyed specific, localised habitats, – all these factors combined to bring about an overwhelming decline.

(*Above*). This rather blurred photo is one of the only photographic images of the ´O´u, a species that was obviously very difficult to capture on film. It has not proved possible to find details of when or how it was taken.

Several of these extinctions happened at the end of the 19th century or start of the 20th, and there is no photographic record of most of the species concerned. In fact, some birds disappeared within just a few years of their discovery and were only ever seen by a handful of ornithologists on a handful of occasions. That there were probably rather more species at this time, and that these never came to attention, seems highly likely.

One species that survived this onslaught of extinction is called the ´O´u, a word pronounced just as might be expected, *Oh-oo.* Its initial escape from oblivion was not to last, however.

(*Above*). A female ´O´u in foliage on Hawaii, photographed in July 1977 by Tim Barr.

A small but plump bird around 17 cm (6–7 inches) long, male ´O´u had a bright yellow head whereas the female was dowdier. This was one of the honeycreepers that evolved a parrot-like bill. In fact its scientific name *Psittirostra* means 'parrot-beak'. However, it was a surprisingly unspecialised feeder and ate fruits and insects, buds and blossom. Nor was it specialised in other ways. Unlike many of its relatives, the ´O´u did not diversify into a different species on each of the main Hawaiian Islands; it remained one species across all of them. This probably indicates that small groups regularly crossed between islands, so populations were never isolated for long enough to develop significant differences.

Able to move between different altitudes, the ´O´u could use varying food sources at different seasons of the year. But when birds descended to lower levels they came into contact with the great scourge of endemic Hawaiian birdlife – the mosquito. These insects spread avian malaria, a disease that proves fatal to honeycreepers, and one that both as individuals and as a species they are wholly unable to withstand.

During the 20th century, ´O´u populations plummeted. By the 1970s the species was almost extinct, with just a few surviving pockets. One of the last reasonably stable colonies lived on the slopes of the volcano, Mauna Loa. During 1984 a lava flow demolished the habitat. The species had a last redoubt on the island of Kaua´i and was reliably seen there in 1988, but a hurricane destroyed this particular area soon after. Hopes are expressed that birds may still survive in places inaccessible to mosquitos, but this becomes increasingly forlorn as time passes. Even if there are a few individuals left, the record of other honeycreeper species hardly inspires confidence that this one could survive for long.

A photograph taken under difficult circumstances by
Robert Shallenberger of one of the last of the ´O´us.
Reproduced courtesy of the photographer.

Mamo
Drepanis pacifica

To make the famous royal feather cloak of Kamehameha I (1758–1819) King of Hawaii, it is estimated that 80,000 Mamos were slaughtered. The cloak is a remarkable object, bright yellow in colour, and can be seen today at the Bernice Bishop Museum in Honolulu. In 1892, many years after it was made, the killing of a single Mamo assumed a very different kind of significance. Just before it died this individual was photographed, and the resulting photo is the only one ever taken of the species. This fact is hardly surprising, for the bird may well have been the last Mamo to die at the hands of humanity, and most of the vast numbers that preceded it died long before the age of photography.

The picture itself is a poor one, yet its story is poignant and comes down to us in some detail, even though more than a hundred years have passed. Walter Rothschild (1868–1937), whose name figures largely in the stories of many extinct birds, was central to the tale. He had sent a collector named Henry Palmer, about whom little is known and who may eventually have been murdered in Australia, to the Hawaiian Islands. His mission was to collect Hawaiian bird specimens for Rothschild's museum, and this mission coincided with a period in which several species stood at the verge of extinction. Nevertheless (or perhaps because of this dire state of affairs), Palmer pursued his objective with relentless application. However, at a crucial point of his expedition he was kicked by a horse and for a while was unable to participate

(*Above*). The photo Ted Wolstenholme (about whom virtually nothing is known) had taken of himself holding his doomed Mamo. The picture was taken in April 1892 on the sides of the volcano Mauna Loa. The photographer was, presumably, Ahulan, an experienced Hawaiian bird catcher who was Wolstenholme's assistant, and the only person known to have been with him at the time. The image is poorly resolved, but its very survival is, in itself, remarkable.

fully in journeys across rugged country. Instead of going himself he sent an employee, Ted Wolstenholme, together with a native bird-catcher called Ahulan on a collecting trip.

They proceeded deep into the forests of Olaa on the side of the volcano, Mauna Loa. Wolstenholme recounted what happened, and his feelings, in a diary that Palmer had instructed him to keep:

> *April 16th 1892. Ahulan fixed the snares and bird-lime … and caught the Mamo! He is a beauty, and takes sugar and water eagerly and roosts on a stick in the tent. I now feel as proud as if someone had sent me two bottles of whisky up.*

Wolstenholme hoped to keep the bird alive, and in the few days it lived with him he became attached to it. At some point the photograph was taken, and after a day or two he returned to where his employer was waiting. Palmer shared none of Wolstenholme's sentimentality. On receiving the bird he promptly killed and skinned it.

The preserved remains ended up in the Rothschild collection, and the stuffed bird that resulted from them was used as one of the models for a hand-coloured lithograph by the accomplished artist J. G. Keulemans (1842–1912). This picture is reproduced on page 231 and shows two birds; the lower one is the individual shown in the photo on page 155.

Some years later, Wolstenholme gave a print of his photograph to George C. Munro (1866–1963), a man who studied Hawaiian birds for many years, during a period when few others were interested. By this means it was preserved for posterity.

Virtually nothing is known of the habits of the species, which was endemic to the island of Hawaii. The Hawaiian hunters who killed birds in their thousands over many generations left little information other than that they were comparatively tame creatures and therefore easy to catch.

It is thought that around 450,000 feathers went into the making of the great cloak, and many more were used to manufacture smaller garments and decorative objects. Although it was the crocus yellow of the feathers that attracted the cloak-makers, these only formed a small part of the plumage, most of which was black. Some writers maintain that when an individual was captured its yellow feathers were ripped off, and the bird was then released. This seems rather unlikely and, in any event, it is barely credible that a small creature like the Mamo (it was 20 cm or 8 inches long) would survive such a violent, shocking and debilitating act. Some have also suggested that hunting was not a primary cause of the species' decline and that these birds were well able to replenish their numbers and make up for any losses. They claim that deforestation and introduced avian disease were the major causes. The likelihood is that all these factors were important elements in the species' demise.

Po´ouli

Melamprosops phaeosoma

One of the characteristics of honeycreepers is the peculiar, musty smell they emit. This smell is so strong and long-lasting it is even detectable on museum specimens preserved more than 100 years ago. The Po´ouli lacked this feature and, along with some structural differences, this caused doubt as to whether it was actually a member of the honeycreeper family. Now, however, it seems to be established that it is one of these birds and systematists generally include it in the group. Curiously, the species had an unexplained attachment to another honeycreeper, the now seriously endangered Maui Parrotbill (*Pseudonestor xanthophrys*), and individuals often followed these other birds around. Obviously, since one species is now extinct and the other nearly so, this relationship can never be investigated.

Quite apart from the smell – or lack of it – and the strange attachment to the parrotbill, the story of the Po´ouli is a remarkable one. The name is a recent Hawaiian invention – meaning 'dark head' – for the local population had no name for this bird, unlike other endemic species. It wasn't discovered until 1973, and there seems to be no tradition of its existence among Hawaiian peoples.

(*Facing page*). A living Po´ouli photographed by Paul Baker after mist-netting the bird during 1997. This individual was released unharmed.

(*Overleaf*). A second photo showing the bird in the hand, taken by Paul Baker.

This sturdy little bird (14 cm or 5.5 inches long) with a short tail was found deep in rainforest at high elevations on the side of the volcano Haleakala, on the island of Maui. At the time of discovery there seemed to be around 200 birds, but in the years that followed this population slumped. By the mid-1990s no more than six or seven individuals were left. Various reasons have been advanced for this catastrophic drop. Avian disease and the decline of a small snail that formed the main food for the species are the best suggestions. During 1997 only three individuals could be found.

In this year a researcher named Paul Baker managed to catch one of these, and it proved to be an adult male. It was caught unharmed in a mist net, and while it was in his care he took a series of very clear photos of it. Before this time the plumage of the adult bird was known only from observations made in the field; no adult bird had been seen up close. The two museum specimens collected and preserved when the species was first discovered were both immature and the plumage was slightly different from that of an adult. After being photographed the bird was released unharmed.

Unfortunately, the three surviving individuals each maintained a distinct home range, and each of these was well away from the others. The chances of the birds coming into contact with one another was, therefore, slim. A scheme to bring them together was devised by conservationists. One of the survivors, a female, was captured and moved to a males' home range (2.5 kilometres or 1.5 miles away) in the hope that they might breed. By the next day, however, something unexpected had happened. The female had flown straight back to her own territory.

Another plan was formulated. It was hoped that all three known birds could be caught and placed in a captive

breeding programme. With such a small gene pool this was
unlikely to prove successful, but it seemed to be the only way
forward. A male bird was captured on September 9th 2004.

(*Above*) Paul Baker took this photo of a Po´ouli in the Hanawi Natural Area, Maui in
1997.

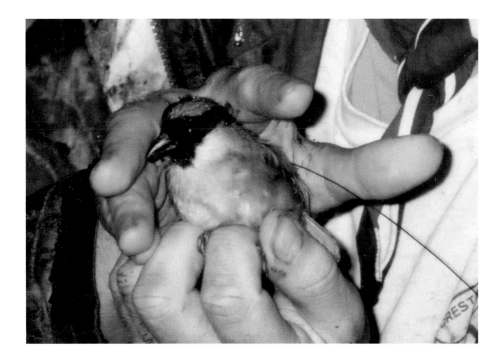

It was old and had only one eye so its survival chances in the wild were limited. So were its chances in captivity. It died on November 24th before a mate could be found. In fact the other two birds went missing and were not seen again.

After the captive male died, some tissue samples were preserved. Alan Lieberman of the San Diego Zoological Society (an organisation that helped to co-ordinate conservation efforts) wrote:

> *Someday, when technology catches up with*
> *our fantasies, we may be able to resurrect the*
> *Po´ouli because we saved the cells.*

(*Above*). This photograph was taken in 2002 as part of the last-ditch attempt to create a pair in the wild, by transferring one of the last three known birds into the territory of another. Photo courtesy of Jim Groombridge.

(*Above*). Another of the series of Po´ouli photos.

Guam Flycatcher
Myiagra freycineti

The fates of extinct creatures are often characterised by grisly circumstances, and that of the Guam Flycatcher is no exception. Guam is a small, isolated island in the Pacific Ocean. As far as the outside world is concerned, its main claim to fame is that it provided the backdrop for one of the vicious confrontations of World War II. Taken by the Japanese just hours after the raid on Pearl Harbour, the island was recaptured after fierce fighting during July 1944. Since then it has returned to more peaceful times but, unfortunately, several of its indigenous animal species have been lost.

The flycatcher, endemic to the island and known there by the name *chuguangguang,* is one of them. It was small (around 13 centimetres or 5 inches long) and males were coloured differently to females. They were glossy blue-black above whereas the females were a brownish grey; both were white below with buff-coloured breasts. Among the more striking features were whiskers around the beak; these helped the birds to locate its insect food.

Often it is human influence on the environment that brings about the extinction of endemic island species. People may destroy the vegetation, hunt the animals, or introduce alien organisms that prey on the indigenous wildlife or compete with it for food. Usually it is rats, cats, dogs, stoats or monkeys that bring about disaster. In this case it was a more unusual introduction: the Brown Tree Snake (*Boiga irregularis*), a species that normally inhabits New Guinea, the

Solomon Islands and parts of Australia. Quite how it was introduced to Guam is uncertain but it seems it hitched rides on American naval vessels soon after World War II. Its arrival had terrible results for several Guam inhabitants. These snakes grow to more than 2 metres (almost 7 feet) in length, and are easily capable of raiding nests placed high in the trees.

The flycatchers were entirely defenceless against this new terror. There was no noticeable reduction in the population until the late 1960s, but the snakes had been steadily establishing themselves, and by this time their presence was becoming more and more obvious.

(*Above*). A photograph taken on Guam in 1981 by Anne F. Maben. This is the male.

The bird's decline was rapid and by 1980 the species was approaching extinction. Early in 1983 birds were recorded in an area known as the Pajon Basin, although there seemed to be fewer than 100. But later in the year only a single individual seemed to be present. Either this bird, or perhaps another that was living nearby, was captured in the hope that a captive breeding programme could be initiated. The bird was a male, but no female was found; in fact no other member of the species could be located. A few months later, on May 15th 1984, this lone male died of unknown causes.

(*Below*). A photo of a Guam Flycatcher nesting in a bamboo clump on Mount Santa Rosa, Guam, taken during the 1940s by an American working in the Pacific war zone.

(*Above*). A photograph featured in a publication titled *The Native Forest Birds of Guam.*
Written by J. Mark Jenkins and published by the American Ornithological Union in
1983. It was taken by the author in 1979 and shows a nest and nestling in a
Casuarina tree.

Thylacine
Thylacinus cynocephalus

The Thylacine is one of the world's most celebrated mystery animals. Does it still exist in remote and little visited parts of Tasmania, or did the last individual die – as the academic record states – on September 7th 1936, in an enclosure at the Beaumaris Zoo, Hobart? The truth probably lies somewhere in between. Small numbers of Thylacines, isolated from others of their kind, were perhaps still padding about the Tasmanian wilderness for some years after the death of the zoo individual that has come to be known as the 'last' of the species. The likelihood is that at some time during the 1940s, 1950s or 1960s the very last Tasmanian Thylacine died alone on a beach, in a forest, or on a mountainside.

Since 1936, many books, pamphlets, magazine pieces and newspaper articles have been written that suggests that Thylacines still survive. Evidence of various kinds has been put forward and comprehensive lists of alleged sightings assembled. Most of these supposed sightings are, for one reason or another, clearly bogus, but some are more convincing. However, as the years go by without concrete evidence coming to light, the odds against survival grow correspondingly longer.

Thylacines, or Tasmanian Tigers as they are often called, were perfectly adapted to their environment. The name 'tiger' is highly misleading. It was given simply because the animal was carnivorous and was striped. In fact, Thylacines looked more like large striped dogs or wolves, a resemblance that is

particularly astonishing when one realises that there is no close relationship between the dog family and Thylacines. Indeed, in zoological terms there is an enormous gulf between them. Despite their dog-like appearance, Thylacines are

(*Above*). A photograph taken at Beaumaris Zoo, Hobart on January 24th 1928 by Benjamin Sheppard. It is often said to be a portrait of the very last captive Thylacine, but this is not the case. This particularly forlorn-looking animal died from disease the day after the photograph was taken. It is probable that it was the source of an infection that resulted in the deaths of at least seven other Thylacines at Beaumaris Zoo that year.

marsupials related to kangaroos and koalas. To put this into context one might say that humans bear a closer relationship to whales than Thylacines do to dogs. Yet if one examines the skull of a Thylacine alongside the skull of a wolf, the similarities are incredible. Apart from certain aspects of the dentition, it is difficult to quickly spot the difference.

(*Facing page*). A photograph taken by a Miss Strickland early in the 20th century at Beaumaris, showing a man feeding a Thylacine.
(*Below*). Another Beaumaris photo, this one taken by an unknown photographer in September 1911. The animal in the foreground does not seem fully adult.

Thylacines once occurred over much of mainland Australia (and also New Guinea), but most zoologists believe that by the time of the coming of Europeans they were restricted to the island of Tasmania. It is thought that they were unable to withstand competition from Dingos (brought to Australia by Aborigines) and were steadily pegged back into remote areas until, eventually, they were eliminated from the mainland altogether.

As Tasmania was settled by Europeans, first as a penal colony then as good farming country, the larger of the indigenous inhabitants were wiped out. First to go was the Tasmanian Emu (*Dromaius novaehollandiae diemenensis*). Next it was the turn of the original human inhabitants.

An elderly lady called Truganini, who died in May 1876, is often said – probably erroneously – to have been the last of her kind.

The Thylacine, being a large carnivore that posed a potential menace to the interests of sheep-rearers, was always likely to come under threat. And so it did. During the 19th century company, governmental and privately funded bounties were paid for the killing of Thylacines.

(*Facing page*). A female Thylacine and her young (around eight months old) at Beaumaris in 1909. Photographer unknown.

(*Above*). The same four animals photographed (possibly by someone named Williamson) at Beaumaris a few months later (January 1910). The animal to the far left was later sold to the Bronx Zoo.

Various amounts were paid; at one time the bounty was £1 for a dead adult, and half that for a pup. The hand of every man with a gun turned against the species. Numbers dwindled. Perhaps the decline was accelerated by some unknown disease (in much the same way as the population of Tasmanian Devils, *Sarcophilus harrisii*, is declining today). By 1900 Thylacines had become very rare. Yet the bounty was still maintained. By curious coincidence it was not until 1936, the supposed year of the species' extinction, that Thylacines were given complete legal protection.

During the last half of the 19th century and on into the 20th, there had been a great demand for living Thylacines from zoos. At London Zoo, for instance, from the 1850s through to the start of World War I, there were almost always Thylacines on show. When one died, another was purchased. There were, therefore, ample opportunities for photography. There are even remarkable film sequences, now easily watched on the internet. Perhaps the best still photos are those taken at the zoo most closely associated with the species, Beaumaris, in Hobart, Tasmania. Pictures were taken there by several people, some of whose names are remembered, some of whom have been long forgotten. Perhaps the most iconic image was taken in 1928 by Benjamin Sheppard. It shows a Thylacine shortly before it died, looking forlornly out of its enclosure (see page 171).

An often-reproduced picture features four Thylacines. It was taken at the Beaumaris Zoo during 1910 but there is uncertainty over who the photographer was. Another Beaumaris photo shows the same four animals a month or two earlier, while another, known to have been taken by a Miss Strickland, shows an adult rearing up against a fence to be fed.

Recently, it has been suggested that DNA could be

extracted from museum specimens and used to recreate the species. Particular publicity has been given to a young animal preserved in spirits at The Australian Museum, Sydney. Whether such an undertaking ever proves possible remains to be seen; it certainly is not a viable course of action within the current state of knowledge and technology.

Most of the attention focused on the possibility of Thylacine survival has, naturally, been directed to the island of Tasmania. Yet equally compelling evidence suggests that the species still lurks on mainland Australia, or even in remote areas of New Guinea where it is known to science from the

(*Below*). There is some doubt about where this photo was taken. It may have been at Beaumaris during the 1920s or at the menagerie of an animal dealer named James Wynyard (*c.* 1916). It is thought to have been taken by Myra B. Sargent.

fossil record. Certain tribes in virtually unexplored areas of Irian Jaya, the western half of the island, speak of a dog-like creature they call the *dobsegna* whose appearance corresponds closely to that of the Thylacine.

The best evidence, however, is in the form of a mummified specimen, one of the treasures of the Western Australian Museum in Perth. This strange exhibit, in a remarkable state of preservation, was found during 1966 in a cave at Mundrabilla Station near the borders of South and Western Australia. The remains were carbon-dated at around 4,500 years old but there is a great aura of mystery surrounding them. A convincing account by Athol Douglas, published in the journal *Cryptozoology* (1990), details reasons why the carbon-dating procedure may have been flawed, and the corpse no more than a few months old when it was found. So far no scientist has cared enough to re-test the specimen and re-examine the evidence.

The last word on the Thylacine might be left to Stephen Sleightholme who produces a comprehensive DVD known as the *International Thylacine Specimen Database*. In a letter to Cameron R. Campbell (who maintains an equally remarkable website called *The Thylacine Museum*) he wrote:

> One must first correct the misconceptions of the ... nineteenth and early part of the twentieth century [when] the scientific community perceived the thylacine as being an evolutionary relic ... primitive and ill-adapted to its island home. This mindset tainted their view of its behaviour. The thylacine was considered ... slow, dumb, stupid, and cowardly, all of which could not be further from the truth. Possibly the reason that so little was written about ... behaviour was due

*to this presumption, the scientific community
taking the view that little could be learnt from the
study of this 'unremarkable' marsupial carnivore.*

(*Above*). The last Thylacine at London Zoo. This animal, a female, was purchased for
£150 during January of 1926 and died on August 9th 1931. The photograph was taken
by F. W. Bond, probably in 1926. This and several other photos reveal the extraordinary
mouth-gape of which these creatures were capable.

Greater Short-tailed Bat
Mystacina robusta

Before the arrival of humans, New Zealand was the land of birds. It is often said that there were no mammals, but this is not quite true. There were a few – some that could swim there and some that could fly. In other words seals and sea lions, cetaceans, and bats. The islands that we now call New Zealand have been isolated from other land masses for many millions of years, and for reasons that are not entirely understood the early mammals that 'inherited' the earth after the fall of the dinosaurs failed to gain dominion here. When the sea cut off these islands from the rest of the world, birds became the dominant life form. Many kinds – including the famous moas and kiwis – gradually lost the power of flight and assumed the roles that are generally taken by mammals. Whether these evolving birds steadily and directly exterminated any early mammals or simply prevented them from successfully competing for ecological niches is not known. Whatever the case, New Zealand became a place entirely uninhabited by terrestrial mammals. Curiously, there may have been an exception to this general rule. During the 19th century rumours surfaced of a strange hairy creature named by the Maori '*waitoreke*'. But no example was ever found by men of science and the '*waitoreke*' – whatever it was – remains an enigma.

(*Facing page*). The only known photo of a Greater Short-tailed Bat. It was taken in 1965, shortly before the species disappeared, by Don Merton one of the heroes of conservation in New Zealand.

180

So, until humans brought cats, dogs, rats, stoats, sheep, cattle, and deer etc., the only mammals that occurred in New Zealand, or visited, were sea-going or flying ones.

As far as bats are concerned, just three species have existed in modern times. One of them, the Long-tailed Bat (*Chalinolobus tuberculatus*) is quite similar to species in Australia and New Caledonia, but the other two are unlike bats found anywhere else in the world. Due to the unusual environmental circumstances that exist in New Zealand they evolved into the world's most terrestrial bats.

In other words, they spend more time on the ground than any of their relatives – they represent the bat family filling New Zealand's vacant 'mouse' niche. The bats scramble over the ground in search of food, with their wing membranes tucked out of the way. Then their arms can be used as front legs, allowing them to run through burrows and forage for food on the forest floor.

The two species are the Greater Short-tailed Bat and the Lesser Short-tailed Bat (*Mystacina tuberculata*) and their terrestrial habits have made them vulnerable to introduced predators and environmental destruction. Although the Lesser Short-tailed Bat still survives, the Greater seems to be extinct. Subfossil remains indicate that it was once widespread over New Zealand, but by the time of European colonisation, or soon after, it was disappearing. It became restricted to Stewart Island in the far south and small islands nearby. At some point the species seems to have vanished from Stewart Island (although it is not impossible that it still survives there) by which time it could only be found on Big South Cape, and a few tiny islands that were close by. These were all sanctuaries for rare New Zealand species and entirely free from introduced mammalian predators. Then, in 1964, rats escaped from a fishing vessel onto Big South Cape, and within a very short space of time wrought havoc among carefully protected island animals, including the bat. The last record of a Greater Short-tailed Bat is of one caught in a mist-net during the April of 1967. Before this, however, an individual was caught and photographed at Puai Cove, Big South Cape Island in 1965 by Don Merton (1939–2011). One of the great figures in the story of the conservation of New Zealand's endangered species.

His photo shows the rather beautiful midnight blue coloration of the bat, and also gives a clear indication of its size.

With creatures as inconspicuous as bats it is, of course, possible that a few might be overlooked. People have reported seeing bats on some of the tiny islands near to Stewart, and bat-like echolocation calls have been recorded.

Caribbean Monk Seal
Monachus tropicalis

In July 1494, during his second voyage to the Americas, Christopher Columbus (1451–1506) landed on a small island to the south of what is now the Dominican Republic. He and his men stayed for three days, and in that time they killed eight seals. Thus began Europeans' relationship with the Caribbean Monk Seal, and it was to carry on much as it had begun, following a similar, but ever-increasing, pattern of destruction for 400 years or so, until the species became rare and finally was exterminated.

It seems that these seals were mostly hunted for the oil that their bodies contained and for their skins, rather than for their meat. As is the case with the Dodo (*Raphus cucullatus*) there are conflicting stories over the desirability of the meat for eating purposes. Most records indicate that it was unpalatable, but some reports suggest it was perfectly acceptable; hungry sailors were, of course, unlikely to be too delicate.

The Caribbean Monk Seal is one of three closely related species, two of which still survive, although both of these are seriously threatened. They are the Mediterranean Monk Seal (*Monachus monachus*) and the Hawaiian Monk Seal (*Monachus schauinslandi*). The 'monk' part of their name comes from their physical appearance. The smooth, round head with rolls of skin around the neck reminded their original scientific describer of a monk dressed in robes. Some have suggested that the name originated due to the seals'

solitary nature, but this is false; in their heyday they could be
seen on land in groups of up to 500.

Historically, the Caribbean Monk Seal came ashore on
isolated islands, keys or atolls, usually surrounded by shallow,
reef-protected waters. Only rarely was it seen on mainland
coasts, but this probably represents a defence against
persecution rather than a natural preference. As its name
suggests, this seal was an inhabitant of the Caribbean and the
Gulf of Mexico.

Hunting reached a climax during the 19th century, and
towards that century's end – as it became apparent that the
species was rare – there was some demand for it from zoos.
However, these animals rarely thrived in captivity; some lived
for only a few days, while the longest known survivor lasted
for five years.

The last reliable sighting of these seals in the wild seems
to have occurred in 1952, when a small colony was seen on

(*Above*). Taken at the New York Aquarium in 1910, this is one of only two photographs
in existence of Caribbean Monk Seals, and shows an adult male. A rope appears to
be tied around the animal's middle. Its purpose is unknown, as is the name of the
photographer.

RARE TROPICAL SEALS.

THE West Indian seals which were received at the Aquarium in June, 1909, still constitute the most noteworthy exhibit in the building.

The possession of three flourishing specimens of a large species near the verge of extinction, is a fact both interesting and important. These seals are the only ones of their kind on exhibition anywhere and may be the last that will ever be seen in captivity.

In the time of Christopher Columbus, this seal was abundant in many parts of the West Indies, its range extending eastward from Yucatan to the Bahamas, Hayti, Cuba and Jamaica. It was gradually exterminated for its oil and skin, and is at the present time known to exist only on the Triangle and Alacran reefs off Yucatan.

The West Indian Seal, (*Monachus tropicalis*), is closely related to *Monachus albiventer* of the Mediterranean, the seal of the ancients, a living specimen of which was exhibited at Marseilles in 1907. Both species are nearly exterminated and with their disappearance the genus *Monachus* will be classed with the extinct animals.

The Aquarium seals will not live forever, and by the time they are gone the man with the gun will more than likely have finished off the remnant of the race in Yucatan. Our seals have not posed to the best advantage for the photographer, but the photographs reproduced in the present BULLETIN, represent so far as we know the only ones in existence of the living animal.

The photographer has been requested to try again, so that the scientist of the future may have all possible documentary evidence as to the general appearance of the animal in life, and its actual existence as late as the year 1910.

These seals, an adult male and two young, are not altogether pleasant as near neighbors, their harsh voices penetrating to every part of the building. The West Indian seal is, so far as our experience goes, the only member of the *Phocidae* or earless seals, that uses its voice in captivity.

The two young seals, a male and a female, have been growing amazingly during the nine months of their life in the Aquarium. They take a fair amount of exercise in the pool, but after being fed usually haul out on the platform along with the large male for a nap, all three huddling close together.

The big male amuses himself occasionally by tossing a flipperful of water in the faces of visitors.

WEST INDIAN SEALS.

Serranilla Bank, approximately halfway between Jamaica and Honduras.

There seem to be just two photos, both taken at the New York Aquarium around 1910. The seals shown in these photographs were captured in either the Mexican state of Campeche or in Yucatan.

Two short articles about them (an adult male and a young male and young female) were published in the *Zoological Society Bulletin* in 1910. The last of these, issued in November of that year, reflects a certain optimism about their condition and adjustment to captive life:

> *The three specimens of the nearly extinct West indian Seals ... which were received by the Aquarium June 14 1909 appear to be in the best of condition. The two younger ones have nearly doubled in size since they came. All three shed their coats during the summer which were quite ragged looking for a time, but are now as sleek as usual. They are fed twice a day on herring and cod, the smaller fishes being fed whole.*

Yangtze River Dolphin
Lipotes vexillifer

The Yangtze River Dolphin has been a subject of enormous interest to the Chinese throughout history and it features in poems, stories, legendary tales, and manuscript texts of a more academic nature. Known as the *baiji* (white dolphin) in China, the species remained largely unknown to western science until 1914, when a 17-year-old American by the name of Charles Hoy shot an individual near Chenglingji. He and his companions ate some of the meat and took a photograph of the dead animal, but the incident would probably have been quickly forgotten except for one thing.

(*Above*). February, 1914. Seventeen year old Charles Hoy with the *baiji* he had just shot, in a water channel connecting the Yangtze with Dongting Lake

188

Hoy cleaned the dolphin's skull and some vertebrae, and took these relics back to the United States, where they ended up at the Smithsonian Institution. There it was realised that these bones belonged to a species unknown to science. During 1918 the dolphin was formally named, and it remains one of the last large animals to be described.

The discovery did not do Hoy much good. He had contracted schistosomiasis from a parasitic flatworm during his time on the Yangtze. This debilitating condition damages the liver and intestines and causes thousands of human deaths, and it left Hoy in a seriously weakened condition. By 1922 he was dead.

There are several river dolphins living in various widely separated parts of the world. They are characterised by their ability to inhabit fresh water, by long thin beaks filled with many small, sharp teeth, and also by poor eyesight; some species are actually blind. The Yangtze Dolphin was found only in the river after which it is named, and various lakes attached to it. Unfortunately for the dolphins (and many other creatures) the area drained by this river system is one of the most densely crowded areas on earth, home to more than ten percent of the world's human population.

There are opposing views as to whether the animal was hunted in historical times. It probably was, but the idea that some local people regarded it as a god, and therefore considered it sacred, has also been suggested. Whatever the case, there is no doubt that the Yangtze River Dolphin still existed in some numbers during the first few decades of the 20th century.

After the communist victory in the civil war in of 1949, various factors combined to ensure the dolphin's decline.

Chairman Mao's 'Great Leap Forward' for the People's Republic led to industrialisation on a vast scale, forest destruction and starvation for millions of people. Naturally, the desperate population turned to any food resources available, and the dolphins suffered.

Then, projects to build huge dams in several places along the Yangtze meant that the dolphin population became fragmented, with local groups separated from others of their kind. So too did populations of fish on which the dolphins fed.

(*Above and facing page*). The famous Qi Qi, photographed at the Wuhan sanctuary where he was held captive for 22 years. He became so used to humans that, apparently, he enjoyed being regularly hauled out of the water.

The whole ecosystem of the river was altered. Other species, may suffer the same fate as the dolphin; a gigantic fish, the Chinese Paddlefish (*Psephurus gladius*), may already have done so. As a result of these changes to its environment, the *baiji* population fell dramatically. Other problems became apparent.

More and more large ships were using the water channels, and the 'white noise' they produced blocked out the auditory ability of the dolphins. Their echolocation systems became less and less effective for hunting and navigation. Many simply crashed into ships' propellers or stranded themselves in their confusion. Massive pollution and new methods of fishing – including blasting and electrifying the waters – spelled the final doom for the species.

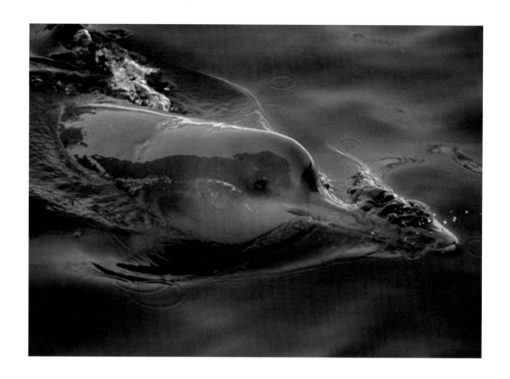

Most photos of the species show the same animal, an individual captured in 1980 and given the name Qi Qi, which was pronounced – although not written – as *chee chee*. While the rest of his kind vanished, Qi Qi became a national celebrity. He was often hoisted out of the water, filmed and then stimulated in forlorn efforts to produce and preserve viable sperm. During 1986 two more *baiji* – a male and a female (who was called Zhen Zhen) – were captured, but after a few weeks the male died despite the desperate efforts of his consort to push him to the water's surface so that he could continue to breathe. Following this death Zhen Zhen was put together with Qi Qi but it seems she hadn't reached sexual maturity and she died before doing so.

Finally, in 2002 Qi Qi died too, apparently of diabetes and old age. He had been in captivity for 22 years, and was given a funeral that was broadcast on national television.

Reports were still coming in of a few wild individuals, but repeated surveys indicated that in reality there were none left. All that remains are a few preserved relics, the photos, and a book by Samuel Turvey (2008) that details the species' story. It is called *Witness to Extinction – How we failed to save the Yangtze River Dolphin*.

(*Facing page*). Almost all existing photos of the Yangtze River Dolphin show the same individual, Qi Qi, in captivity at Wuhan. These two were taken in 1988.

Quagga
Equus quagga quagga

The Quagga is one of the icons of extinction with a very distinct identity, yet recent research has revealed that it is probably not even a full species – just a race of the still-extant Plains Zebra; DNA analysis seems to confirm this. Despite wide acceptance of this idea by biologists there is little doubt that the Quagga will retain its iconic position, its dramatic history and unmistakable appearance ensuring that this status will endure.

Although shaped like any other form of zebra, Quaggas were startlingly different in coloration. They only carried stripes on the head, neck and front of the body, and most of the areas that are white in typical zebras were brown, with only the legs and underparts white.

One of the reasons for the Quagga's enduring celebrity is surely its peculiar sounding name. Like an even more famous extinct creature, the Dodo (*Raphus cucullatus*), it carries a name that is short but unforgettable. In fact *quagga* is derived from a Hottentot expression that represented an attempt to imitate the animal's barking call.

Curiously, the name caused confusion long before the DNA evidence came along, because at one time Europeans applied it to all zebra-like creatures. Only later did it become restricted to those animals we now describe as Quaggas. In many respects the scientific debate over the status of the Quagga reflects the whole issue of exactly what constitutes full species, and what does not. Although there are clearly

(*Above*). There are five known photographs of the Quagga in existence. All are pictures of the same individual, a female that was a resident at London Zoo, although they are the work of three different photographers. This rather blurred image is the least well-known and was shot by an unknown photographer in the 1860s.

set-down explanations of how the word 'species' should be defined these are ultimately all open to interpretation. In other words, one biologist might interpret the evidence in a far stricter way than another. Whereas no-one would suggest that a Lion belongs to the same species as a Tiger, it is rather more problematic to determine whether the Asiatic Lion is a species distinct from the Lions that inhabit Africa; most scientists accept that they do form a single species, but some might argue!

And so it is with zebras. The majority of authorities divide them into just three species, but in the past these three were split up into several more. This discrepancy causes serious confusion in the matter of scientific naming, and what should be a most precise method of identification becomes something of an intellectual minefield that sometimes takes a good deal of decoding.

The Plains Zebra was once known scientifically as *Equus burchelli* and the Quagga as *Equus quagga*, but when the DNA evidence seemed to reveal that both belonged to the same species the rules of zoological nomenclature had to be applied, and the names re-aligned. The oldest published name (no matter how confusing or inappropriate) must be given priority, and the oldest is *Equus quagga* (being first used in 1778). The Plains Zebra therefore became *Equus quagga burchelli* and the Quagga became *Equus quagga quagga*. And to make for even greater confusion, there is the problem of the group of animals once described as Burchell's Zebras. These were considered to constitute an entirely separate species to other plains zebras. Photos taken in the 19th century showing captive animals

(*Above*). Taken by Frederick York during the summer of 1870, this is the best of the photographs showing London Zoo's female Quagga.

with entirely unstriped legs were regarded as proof that these creatures were quite distinct, and that they became extinct at some time during World War I. However, in recent years it has been decided that these individuals showed no fundamental, repeatable difference to other plains zebras, and therefore their kind is not extinct at all.

In terms of the Quagga itself, none of this arcane complication matters very much, for the concept of Quaggas as separate, clearly defined animals still retains popularity and is grounded in their well-established historical profile.

During the 18th and 19th centuries Quaggas were very common inhabitants of the grassy plains of drier areas of southern Africa. Like other zebras they were not particularly easy to tame or put to domestic use, and since they were regarded as a pest that competed with cattle and sheep for grazing, they were regularly hunted for their meat and their hides. This hunting became so intensive that it was eventually banned. The restriction happened in 1886, but it was far too late. The last known Quagga was already dead. It had expired at the Amsterdam Zoo three years previously. The last wild individual had probably been shot during the 1870s.

There are just five known photographs of a living Quagga, all of them featuring the same individual, a female resident at London Zoo. She died on July 15th, 1872, having been at the zoo for 21 years. Her body was passed to a firm of taxidermists and osteologists named Gerrards, who had premises close to the zoo and therefore gained access to many of the animals

(*Facing page*). Burchell's Zebra. This photograph, taken by T. J. Dixon around the year 1885 shows an individual in captivity at London Zoo. It reveals clearly the unstriped legs that were characteristic of this form of Plains Zebra.

(*Above*). This photograph was taken by Frank Haes in 1864. It was commercially produced to be looked at through a device known as a stereoscopic viewer, which gave an illusion of a 3D image.

(*Above*). An image from the 1860s by an unknown photographer, possibly Frank Haes.

that died there. Gerrards prepared both the Quagga's skin and her skeleton. The skeleton was sold for the sum of £10 to the famous American dinosaur hunter Othniel C. Marsh (1831–1899), and it is now in the Peabody Museum at Yale University. The stuffed skin was eventually sold to the Royal Scottish Museum, Edinburgh.

At least one of the photos was taken by a gentleman called Frank Haes (1832–1916), who took many pictures of zoo animals and sold the results of his work in the form of stereoscopic cards. On these, the photo was printed twice (side by side) and when looked at through a gadget known as a stereoscopic viewer a startling 3D image was obtained. Haes left some interesting remarks concerning the process by which he obtained his pictures in an article published in *The Photographic Journal* for January 16th, 1865. He would often spend 20 minutes or so in coaxing or threatening his subject into an appropriate position, and then take his shot. Often the effort was wasted because the wet plate (then a very necessary part of one's photographic equipment) would dry out before he and the animal were ready. Then there was another problem associated with speed. Exposures were made as quickly as possible, but this still took around a third of a second and any movement on the part of the animal was likely to spoil things.

The two best images were taken by Frederick York (1823–1903) who, like Haes, specialised in stereoscopic views, but who also produced many images as slides for the magic lantern, another 19th-century device designed to enhance viewing pleasure.

Perhaps the most important figure in the history of Quagga research was the South African naturalist Reinhold Rau (1932–2006). Rau took tissue samples from stuffed

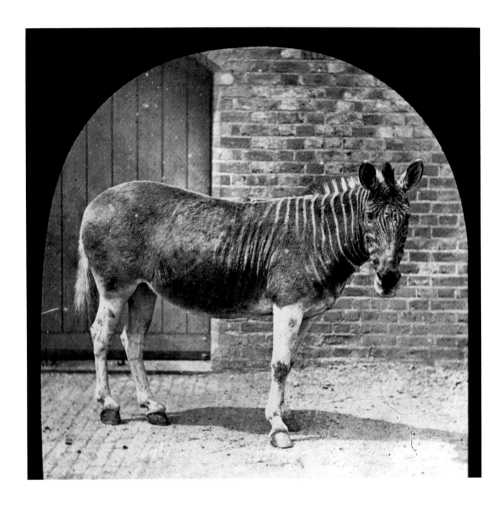

museum specimens. One in particular had been poorly produced, and he was able to extract muscle and tissue that in ordinary circumstances would have been removed by the taxidermist during the process of skin preservation. It was the results of an analysis of these samples that convinced him that the Quagga was not a distinct species, but simply a race of

(*Above*). A reproduction of a silver gelatin print by Frederick York that was also used as a magic lantern slide. This picture was probably taken during the summer of 1870.

the Plains Zebra. Following examination of the evidence, he formed the opinion that it would be possible to re-create Quaggas, and with this end in mind he pioneered a course of action that has become known as The Quagga Project. In recent years selective breeding has begun to achieve the results that Rau hoped for. To those who questioned the conceptual basis of his programme, he had a simple response:

> *The Quagga is a Quagga because of the way it looked, and if you produce animals* [from selective breeding] *that look that way, then they are Quaggas.*

(*Facing page*). A zebra captured in 1993 that had Quagga-like features. It was subsequently named Howey, and used in the selective breeding programme that may lead to the 'resurrection' of these animals.

Schomburgk's Deer
Cervus schomburgki

(*Above*). A photograph taken in 1911 at a Berlin zoo showing a captive Schomburgk's deer. Photographer unknown.

Despite the fact that it was once quite common, Schomburgk's Deer is a mysterious animal. There appears to be only one preserved specimen in the world's museums, a stuffed animal in the collection of the Natural History Museum, Paris. Similarly, there seems to be just a single photograph, and this was taken at a Berlin zoo during 1911 (another purporting to show Schomburgk's deer may actually show another species). In fact very few individuals were kept in zoos; it seems that no more than seven reached Europe and not one ever appeared in North America. There is no record of a European ever having seen a living example in the wilds of Thailand – the species' stronghold – or any other part of south-east Asia. Despite this, there are perhaps as many as 400 sets of antlers still preserved in museums or in private hands, and such antlers once figured strongly in the Chinese pharmaceutical trade. The best surviving ones show that they were very elaborately branched and could boast a surprising number of points.

The species was first described in 1863 and was named after Sir Robert Schomburgk, who was the British consul in Bangkok. It inhabited swampy plains with growths of bamboo and long grass – mostly in Thailand, but perhaps also in neighbouring countries. These deer avoided areas of really dense vegetation but were plentiful in the places they favoured. Standing around a metre (3.5 feet) high at the shoulder, with rich chocolate-brown colouring and spectacular antlers, they made an attractive target for hunters. During times of flooding the small herds in which the animals generally lived were often forced to crowd onto higher ground and these high points sometimes became 'islands' that could easily be surrounded by humans with guns or other weapons.

The inevitable massacres followed. However, it was probably not this that caused the species' extinction. Increasingly large-scale production of rice led to the destruction of the very areas that the deer inhabited, and the species became rare as the 19th century passed into the 20th.

As far as is known, the animals were extinct in the wild by the early 1930s. One individual – perhaps the last – was kept as a pet in a temple in the Samut Sakhon province of Thailand, where it lived until 1938. There is a story that it was killed in that year by a local drunk.

However, this may not actually have been the last Schomburgk's Deer. In February 1991 a set of antlers was seen at a Chinese pharmacy in Laos and photographed. These pharmacies are often full of products from the natural world, most notably Tiger bones and ground-up rhino horn. Among such relics there are often pieces of antler, but it is claimed that this particular set came from an individual killed nearby the year before, and that this individual was a Schomburgk's Deer. If this is so, then clearly the species was still in existence in the early 1990s – but is identification of a set of antlers from a single photo taken in a Chinese pharmacy completely convincing?

(*Facing page*). A photograph from an unknown source that may, or may not, show Schomburgk's Deer.

Bubal Hartebeest
Alcelaphus buselaphus

The question of whether or not the Bubal Hartebeest constitutes a full species is something of a vexed one. Many authorities consider it to be a race of a still common species, while others believe it to be quite distinct. Whatever is ultimately decided, this form did have a well-defined geographical range and a clear historical identity.

In fact, the strange name *hartebeest* is applied to several closely related species that occur widely across Africa. It is an amalgam of two Afrikaans words, *hert* meaning deer and *beest* meaning (as might be expected) beast. Clearly the Boer settlers who provided the name weren't quite sure how to categorise these rather peculiar-looking animals. The word *bubal*, which is only applied to this particular hartebeest, seems to derive from Greek and means 'gazelle' or 'ox'.

Bubals were the only kind of hartebeest that lived north of the Sahara; the others are all strictly sub-Saharan. Once, they ranged from Algeria and Morocco in the west to Egypt in the east. Because of the geography of this distribution Bubal Hartebeests were well-known to the ancients. Their horns have been found in Egyptian tombs, giving rise to the idea that these animals may have been domesticated, and also used for sacrificial purposes. There are depictions of them on Roman mosaics at Hippo Regius (the modern city of Annaba) in Algeria, and they are mentioned by Aristotle, Aeschylus and Pliny. There even seems to be a reference in the Old Testament (I Kings 4:23) where they go by the name of *yachmur*, giving rise to the suggestion that they may have lived in Palestine.

Towards the end of the 19th century and on into early years of the 20th, several of these hartebeests were kept in zoos, but only one seems to have been photographed.

(*Above*). A female Bubal Hartebeest photographed around 1895 at London Zoo by Lewis Medland.

This was a female that lived at London Zoo from 1883 until 1897. The picture was taken by Lewis Medland (1845–1914), probably during 1895, and the same individual served as the model for a painting by the famous wildlife painter Joseph Wolf (1820–1899) that was copied as a lithograph (see page 235) by his friend Joseph Smit (1836–1929) and published in a celebrated four-volume tome called *The Book of Antelopes* (1894–1900) by P. L. Sclater and Oldfield Thomas.

Another individual was at the same zoo for a few months during 1906 and 1907, but it appears that it did not have its portrait taken. Nor, it seems, did one that was at the Jardin des Plantes in Paris, and this animal may well have been the last of its kind. It died on November 9th, 1923.

After this there are only unsubstantiated rumours that a few of these hartebeests lingered in the wild. An individual may have been shot in Algeria or Morocco in 1925. Others may have been sighted in remote areas of the Atlas Mountains.

Although it was common in historical times, by the 19th century the range of the Bubal Hartebeest had been pinned back to Morocco and Algeria. Originally, its primary enemy seems to have been the Barbary Lion, a race of Lion that is itself now extinct in the wild. Humans became a rather more unrelenting foe, and our reign of terror culminated in 19th century massacres when the French military occupying Morocco and Algeria mowed down herds for sport and food.

There is a record of what may have been the last colony of wild Bubal Hartebeests, which was found in 1917 somewhere in the Atlas Mountains. There were 15 of them.

All but three were slaughtered by a single hunter.

Appendix

Some of the photographs reproduced in this book are certainly inadequate as clear representations of the species they show. Although fascinating as historical records, they may leave the viewer frustrated, with a wish to visualise more exactly what the species actually looked like. Sometimes this may simply be because the image is in black-and-white and gives no impression of colour; on other occasions it is because it is too blurred or too distant. The photo of a Mamo (see page 155) is a good example. It is blurred, was shot in black-and-white and the subject is little more than a vague image. But there is no other available photograph of the living bird.

This Appendix exists to make good these kind of defects, in as illuminating a way as possible. Reproductions of paintings are used to show characteristics of species that are not evident in the photographs. Some of these paintings were produced at times when the species themselves were still extant; where no such image exists, more modern paintings are featured.

For some species the photos used in this book are reasonably clear or it is felt that a coloured picture would not add significant information. Such species are therefore excluded from this appendix. These excluded species are:

Atitlán Giant Grebe   Greater Short-tailed Bat
Aldabra Brush Warbler   Caribbean Monk Seal
Po´ouli   Yangtze River Dolphin
Guam Flycatcher   Schomburgk's Deer

(*Above*). Alaotra Grebe. Oils on canvas (*c.* 2009). Chris Rose. Private collection. Courtesy of the artist.

(*Above*). Pink-headed Duck. Watercolour (*c.* 1780). Musavir Bhawani Das. Liverpool Museum, Lady Impey Collection.

(*Above*). Heath Hens (female left, male in display right). Watercolour (*c.* 1926). Louis Agassiz Fuertes (1874–1927). This illustration is taken from *The Heath Hen* (1928) by Alfred O. Gross.

(*Facing page, above*). Wake Island Rail. Oils on canvas (1986). Errol Fuller. Private collection.

(*Facing page, below*). Laysan Rail and albatross. Acrylic on paper (1999). Julian Pender Hume. Private collection. Courtesy of the artist.

(*Above*). Eskimo Curlews. Hand-coloured lithograph. J. G. Keulemans (1842–1912). This illustration is taken from *A History of the Birds of Europe* (1871–1881) by Henry Eeles Dresser.

(*Above*). Passenger Pigeons. Acrylic on paper (2010). Julian Pender Hume. Private collection. This painting is loosely based on a well-known 19th century engraving showing a Passenger Pigeon hunt. Courtesy of the artist.

(*Above*). Carolina Parakeet. Watercolour (*c.* 1801). Jacques Barraband (1768–1809). Private collection. An image engraved after this painting was reproduced in *Histoire Naturelle des Perroquets* (1801–1805) by Francois Levaillant.

(*Above*). Paradise Parrots (male above, female below). Watercolour (1979). William T. Cooper. Private collection. This watercolour was reproduced in *Australian Parrots* (1981) by J. M. Forshaw and William T. Cooper. Courtesy of the artist.

(*Facing page*). Laughing Owls. Hand-coloured lithograph. J. G. Keulemans (1842–1912). This illustration is taken from *Ornithological Miscellany* (1875–1878) by George Dawson Rowley.

(*Above*). Ivory-billed Woodpeckers (male left, females right). Aquatint by John James Audubon (1785–1851) and Robert Havell (1793–1878) based on a watercolour by Audubon. This illustration is taken from *The Birds of America* (1827–1838) by John James Audubon.

(*Above*). Imperial Woodpeckers (male left, female right). Watercolour (*c.* 1965). George Sandstrom (1925–2006). Private collection. Physical differences between this species and the last are slight, the main one being that the white on the shoulder is narrower and does not reach the face. Courtesy of the artist.

(*Above*). New Zealand Bush Wrens (adult above, immature below). Hand-coloured lithograph. J. G. Keulemans (1842–1912). This illustration is taken from the supplement to *A History of the Birds of New Zealand* (1905) by Sir Walter Buller.

(*Above*). Bachman's Warblers (male above, female below). Watercolour (*c.* 1833). John James Audubon (1785–1851). New York Historical Society.

(*Facing page*). Kauaʻi ʻOʻos (immature above, adult below).
(*Above*). Mamos. The individual in the photo on page 155 served as a model for the lower of these two figures.
Both pictures are hand-coloured lithographs by J. G. Keulemans (1842–1912) and are taken from *Avifauna of Laysan* (1893–1900) by Walter Rothschild.

(*Above*). ´O´us (female above, male below). Hand-coloured lithograph.
F. W. Frohawk (1861–1946). This illustration is taken from *Aves Hawaiienses*
(1890–1899) by S. Wilson and A. H. Evans.

(*Above*). Thylacines. Hand-coloured lithograph. Henry Constantine Richter (1821–1902). This illustration is taken from *Mammals of Australia* (1845–1863) by John Gould.

(*Above*). Quagga. Oils on canvas (*c.* 1821). Jacques-Laurent Agasse (1767–1849). Hunterian Collection, Royal College of Surgeons, London.

(*Above*). Bubal Hartebeests. Hand-coloured lithograph. Joseph Smit (1836–1929) and Joseph Wolf (1820-1899). This illustration is taken from *The Book of Antelopes* (1894–1900) by P. L. Sclater and Oldfield Thomas.

Further Reading

Numerous books on specific subjects are also referenced in relevant places in the text. Some (*Aves Hawaiienses* by S. Wilson and A. H. Evans, for instance), are rare or beyond the means of most readers, but are available for inspection on the internet.

Allen, Glover M. 1942. *Extinct and Vanishing Mammals of the Western Hemisphere.* New York.

Bales, Stephen Lyn. 2010. *Ghost Birds.* Knoxville.

Barnaby, David. 1996. *Quaggas and other Zebras.* Plymouth.

Bodsworth, Fred. 1955. *Last of the Curlews.* London.

Buller, Walter. 1888–9. *A History of the Birds of New Zealand.* London.

Coues, Elliot. 1874. *Birds of the North West.* Washington.

Cokinos, Christopher. 2000. *Hope Is the Thing with Feathers.* New York.

Collar, N. Gonzaga, L., et al. 1992. *Threatened Birds of the Americas, The ICBP/IUCN Red Data Book.* Cambridge.

Edwards, John. 1996. *London Zoo from Old Photographs, 1852–1914.* London.

Fuller, Errol. 1999. *The Great Auk.* Southborough, Kent.

Fuller, Errol. 2001. *Extinct Birds.* Oxford and Ithaca.

Gallagher, Tim. 2005. *The Grail Bird – Hot on the Trail of the Ivory-billed Woodpecker.* Boston.

Gallagher, Tim. 2013. *Imperial Dreams: Tracking the Imperial Woodpecker through the Wild Sierra Madre.* New York.

Gollop, J. Barry, T. and Iversen, E. 1986. *Eskimo Curlew, A Vanishing Species.* Regina Saskatchewan.

Greenway, John. 1958. *Extinct and Vanishing Birds of the World.* New York.

Grooch, William. 1936. *Skyway to Asia.* New York.

Gross, Alfred. 1928. *The Heath Hen.* Boston.

Guiler, Eric. 1985. *Thylacine: The Tragedy of the Tasmanian Tiger.* Melbourne.

Guthrie-Smith, Herbert. 1925. *Bird Life on Island and Shore.* London.

Harper, Francis. 1945. *Extinct and Vanishing Mammals of the Old World.* New York.

Hirschfeld, Eric, Swash, Andy and Still, Robert. 2013. *The World's Rarest Birds.* Princeton.

Hume, Julian P. and Michael Walters. 2012. *Extinct Birds.* London.

IUCN. 2011. *Species on the Edge of Survival.* Gland, Switzerland.

Jenkins, J. Mark. 1983. *The Native Forest Birds of Guam.* Washington.

Knox, Alan and Walters, Michael. 1994. *Extinct and Endangered Birds in the Collections of the Natural History Museum.* London.

LaBastille, Anne. 1990. *Mama Poc.* New York.

Milner, Richard. 2009. *Darwin's Universe, Evolution from A – Z.* Los Angeles.

Mittelbach, Margaret and Crewdson, Michael. 2005. *Carnivorous Nights. On the Trail of the Tasmanian Tiger.* Edinburgh.

Morris, Rod and Smith, Hal. 1988. *Wild South. Saving New Zealand's Endangered Birds.* Auckland.

Nowak, Ronald. 1999. *Walker's Mammals of the World.* Baltimore.

Olsen, Penny. 2007. *Glimpses of Paradise. The Quest for the Beautiful Parakeet.* Canberra.

Pratt, H. Douglas. 2005. *The Hawaiian Honeycreepers.* Oxford.

Quammen, David. 1996. *The Song of the Dodo.* London.

Rothschild, Miriam. 1983. *Dear Lord Rothschild.* London.

Rothschild, Walter. 1893-1900. *The Avifauna of Laysan and the Neighbouring Islands.* London.

Rothschild, Walter. 1907. *Extinct Birds.* London.

Schorger, A. 1955. *The Passenger Pigeon.* Madison, Wisconsin.

Shuker, Karl. 1993. *The Lost Ark.* London.

Snyder, Noel. 2004. *The Carolina Parakeet: Glimpses of a Vanished Bird.* Princeton.

Snyder, Noel. 2009. *The Travails of Two Woodpeckers: Ivory-bills and Imperials.* Albuquerque.

Stattersfield, A. and Capper, D. 2000. *Threatened Birds of the World.* Cambridge.

Tanner, James. 1942. *The Ivory-billed Woodpecker.* New York.

Thornback, Jane and Jenkins, Martin. 1982. *The IUCN Mammal Red Data Book.* Gland, Switzerland and Cambridge.

Turvey, Samuel. 2008. *Witness to Extinction. How We Failed to Save the Yangtze River Dolphin.* Oxford.

Wilson, Scott and Evans, Arthur Humble. 1890–1899. *Aves Hawaiienses.* London.

(*Overleaf*). Benjamin, the last captive Thylacine, photographed by the celebrated Australian naturalist David Fleay (1907–1993) at Beaumaris Zoo, Hobart on December 19th, 1933. There is some doubt as to whether the name 'Benjamin' was ever given to the animal during its lifetime but, whatever the truth, the name has stuck and become part of Thylacine legend. Similarly, the sex of the animal has sometimes been questioned, but Thylacine expert Stephen Sleightholme has established beyond doubt that it was a male. The remarkable ability of the species to open its jaws so wide has often been commented on (in fact it could achieve the largest angle of 'gape' of any known mammal), but the precise reason for this capacity is not understood. However, it is known that while Mr Fleay was lining up and focusing the lens on his large Graflex camera, Benjamin bit him on the bottom!

Acknowledgements

The author would like to thank all the people who have kindly given access to images featured in this book. In particular he would like to thank Nancy Tanner whose remarkable contribution has been so important. Also special thanks should be given to Stephen Lyn Bales, Julian Hume, Chris Rose, William T. Cooper, Don Merton, Robert Schallenburger, H. Douglas Pratt, Madeleine Thompson, John Edwards, Robert Prŷs-Jones, Martin Lammertink, Frank S. Todd, Nigel Collar, Dr. Stephen Sleightholme, Richard Thorns, Paul Thompson, Rosemary Fleay-Thompson and Stephen Fleay. If anyone feels that a photographic copyright has been infringed they are invited to contact the author, care of the publishers.

(*Overleaf*). Was this the last Thylacine? Perhaps. Benjamin, photographed at the Beaumaris Zoo, Hobart shortly before his death on September 7th, 1936. The photographer is unknown.

Index

Page numbers in *italic* denote an image. Those in **bold** denote a chapter.

(*Overleaf*). The earliest known photograph of a Thylacine (in fact the only photo of a live animal taken in the 19th century). It is one of a series of zoo animals photographed in 1864 by Frank Haes and published as 'stereoscopic views' (see page 202), mounted on yellow card. Although the series was widely distributed, only two examples of this particular card are known to have survived. Concerning this photo, Haes wrote:

> It was necessary to go into the enclosure: these animals are savage, cowardly and treacherous; and a pretty dance they led us. The keeper despaired of ever getting one of them as we required … one did his best to catch hold of my legs; however, we at last tired one out and fixed him on a plate, though the position is not all that could be wished.

The skeleton of this forlorn male animal was later preserved, and is now in the Royal Scottish Museum, Edinburgh. Reproduced courtesy of John Edwards.

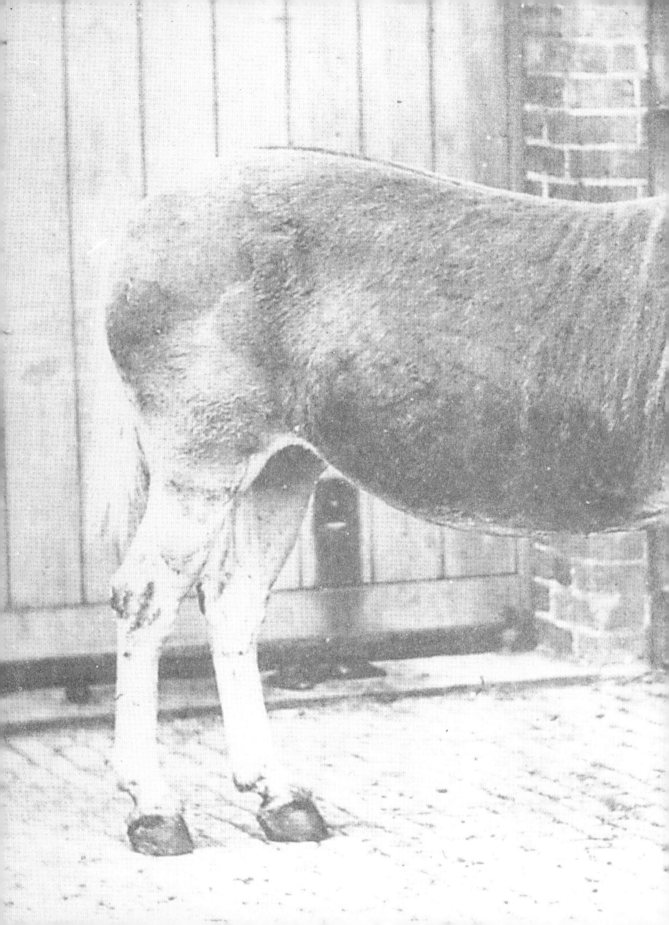